Fatigue, Safety and the Truck Driver

Fatigue, Safety and the Truck Driver

Nicholas McDonald

Trinity College, Dublin

Taylor & Francis
London and Philadelphia
1984

UK   Taylor & Francis Ltd, 4 John St., London WC1N 2ET

USA   Taylor & Francis Inc., 242 Cherry St., Philadelphia, PA 19106-1906

British Library Cataloguing in Publication Data

McDonald, Nicholas
Fatigue, safety and the truck driver.
1. Motor-truck drivers   2. Fatigue
I. Title
388.3'24044   TL230.3
ISBN 0-85066-207-9

Library of Congress Cataloging in Publication Data

McDonald, Nicholas.
Fatigue, safety and the truck driver.
Bibliography: p.
Includes index.
1. Truck drivers—Diseases and hygiene.   2. Fatigue.
3. Truck driving—Physiological aspects.   4. Truck
driving—Safety measures. I. Title.
RC1032.M37 1984   363.1'19388324   83–18185
ISBN 0–85066–207–9

Printed in Great Britain by Taylor & Francis (Printers) Ltd, Basingstoke

Preface

This book is concerned with the conditions which might give rise to fatigue among truck drivers, and the effects of fatigue on driving safety. This might seem to be a straightforward undertaking leading to fairly clear-cut guidelines for practice. That this is not so is partly due to an incomplete matching of the demands of 'good scientific practice' and the requirement that the findings of research should be practical and applicable.

In general, scientists look primarily to their peers in the scientific community for recognition and evaluation of their work. The practical relevance or applicability of their work is not usually an issue. When the topic of research concerns an issue of practice or policy within the world of everyday social activity, the scientist's task, at first glance, would seem to be one of subjecting a set of propositions derived from the relevant domain of everyday practice to systematic and rigorous empirical test; then making a definite policy recommendation on the basis of the outcome. However, it is no easy matter to formulate a fairly complex social problem as a set of testable hypotheses and operational procedures that have good 'ecological validity'. The interpretation of the measures obtained provides a subsequent and no less difficult problem.

What frequently happens is that an area of research tends to generate an independent momentum of its own—a number of empirical investigations are conducted which in turn present subsequent problems of clarification of hypotheses, refinement of techniques, and theoretical evaluation of results. What may have begun as a relatively practical problem being fairly directly tested can easily generate an (almost self-perpetuating) industry of scientific activity which can be undertaken with frequently very little regard to the nexus of social problems that gave rise to the question in the first place. At some time it becomes necessary to take stock in a critical and evaluative manner of the achievements of an accumulated body of research findings. This book is an attempt to do so for the topic of fatigue and driving safety, in terms both of the scientific adequacy and theoretical integration of the accumulated work, as well as the implications of this work for practice.

Numerous people have helped in the process of compilation and sifting of the evidence and formulating ideas, but none more than Jack Sandover of Loughborough University and Ray Fuller of Trinity College, Dublin. Paul Thomas and Tony Kaye were stimulating colleagues on a research team investigating this topic. I am also grateful for the comments and encouragement of the late Professor J. Weiner of the M.R.C. Environmental Physiology Unit at the London School of Hygiene and Tropical Medicine, and Ivan Brown of the M.R.C. Applied Psychology Unit at Cambridge. I also acknowledge the support of the Medical Research Council and the Transport and Road Research Laboratory in funding the research and allowing it to be published. I also appreciate the efforts of those who have typed the various stages of the manuscript, particularly Peggy Brennan.

I would like to thank the following who have kindly given permission for the use of copyright material: Hodder and Stoughton Educational for a figure from "The sleep of train drivers" by J. Foret and G. Lantin in *Aspects of Human Efficiency* (W. P. Colquhoun, ed.). Academic Press for figures from "Noise in heavy goods vehicles" by D. Williams and W. Tempest, 1975, *Journal of Sound and Vibration*, **43**, 97–107, and from "A review of the effects of vibration on visual acuity and continuous manual control Part I" by M. J. Griffin and C. H. Lewis, 1978, *Journal of Sound and Vibration*, **56**, 383–413; both copyright by Academic Press (London) Ltd. John Wiley and Sons for a figure from "12 and 24 hour rhythms in error frequency of locomotive drivers and the influence of tiredness" by G. Hildebrandt, W. Rohmert and J. Rutenfranz. 1974, *International Journal of Chronobiology*, **2**, 175–180. Westdeutscher Verlag GmbH for two figures from "The influence of fatigue and rest period on the circadian variation of error frequency in shift workers" by G. Hildebrandt, W. Rohmert and J. Rutenfranz, in *Experimental Studies of Shiftwork* (W. P. Colquhoun, ed.). The Transportation Research Board, National Academy of Sciences, Washington, D.C. for a figure from "Driver eye movement patterns under conditions of prolonged driving and sleep deprivation" by N. A. Kaluger and G. L. Smith, Jr, *Highway Research Record*, **336**, 92–106. North-Holland Publishing Company for a figure from *Road-User Behaviour and Traffic Accidents* by R. Näätänen and H. Summala. Human Factors Research, Inc., the National Highway Traffic Safety Administration and the Bureau of Motor Carrier Safety for figures from *A Study of the Relationships among Fatigue, Hours of Service and Safety of Operations of Truck and Bus Drivers* by W. Harris, R. R. Mackie, C. Abrams, D. N. Buckner, A. Harabedian, J. F. O'Hanlon and J. R. Starks; and from *A Study of Heat, Noise and Vibration in Relation to Driver Performance and Physiological Status* by R. R. Mackie, J. F. O'Hanlon and M. E. McAuley. Routledge and Kegan Paul Ltd for a figure from *The Lorry Driver* by P. G. Hollowell (1968). The Society of Automotive Engineers for figures from *How Loud are Diesel Truck Cabs?* by J. D. Hutton, SAE Paper no. 720698; and from *A Study of Vehicle*

Vibration Spectra as Related to Seating Dynamics by L. F. Stikeleather, G. O. Hall and A. O. Radke, SAE Paper no. 720001.

Every effort has been made to trace all the copyright holders but if any have been inadvertently overlooked the publishers will be pleased to make the necessary arrangements at the first opportunity.

Nicholas McDonald
Dublin
October 1983

Contents

To Ann, Dick and Dorothy

Chapter 1
Introduction

In comparison with its assumed importance in the public consciousness the actual evidence that fatigue makes an important contribution to road accidents is by no means clear cut, only partial and frequently indirect. The following chapters represent an attempt to assess this evidence in a comprehensive manner, drawing conclusions wherever the evidence permits and pointing out the short-comings in the pattern of evidence where the inferences that may be drawn are only tentative or where no inferences are warranted. This analysis of the literature takes as its point of reference the work of heavy goods vehicle (HGV) drivers, though much of what is said is equally applicable to other professional drivers and the general driving population; and indeed much of the evidence, particularly the experimental evidence, derives from non-professional drivers of cars. The importance of safety in the haulage industry is underlined by the number of workers potentially affected: in Great Britain there are between 900 000 and one million holders of HGV licences (Department of the Environment 1980). In a wider context it has been estimated that there are 55 million people employed in the road transport industry worldwide, with 15 million in Europe and 8 million in the USA (ILO 1977).

1.1. The changing structure of the haulage industry

Since World War II there has been a massive expansion in the road haulage industry in both developed and developing countries. This expansion has brought about considerable changes in the structure of the industry which in turn have radical implications for the patterns of work of the drivers themselves. These changes are discussed by McDonald (1980).

To illustrate this expansion, employment in haulage companies in the USA more than doubled between 1947 and 1974 (while employment in railway companies was cut to between one-half and one-third of the 1947 level). Both the number and the size of firms has increased: between 1951 and 1971 the number of haulage companies went up from 47 000 to 65 000, and the number employing over 500 employees increased from 34 to 88 (Friedman 1978). Table 1.1 shows the development of increasing size of seven major trucking companies in the USA. The composition of the haulage fleet has also changed: there are more trucks, but the biggest expansion has been in the number of the heaviest vehicles. Thus, in Britain in the 20 years up to 1980 the number of trucks over eight tons unladen weight increased over ten-fold (Ministry of Transport 1967, Department of Transport 1981). The mileage performed by these vehicles has also increased very much more than the numbers of vehicles, and the greatest mileages are performed by the heaviest trucks. Productivity in the industry in terms of ton miles of freight shifted has thus mushroomed and, in value terms, it has been estimated (for the USA) that output per man-hour in road freight transport has risen more rapidly than for the private domestic economy as a whole during the post-war years (Felton 1978).

Increasing productivity in the haulage industry has not just come from the increasing size, weight and mileage of vehicles but from the introduction of more sophisticated management techniques which emphasize cost cutting and efficiency, for example by the computer planning of the loading, dispatching and routing of trucks, and the redesign of terminals. But more importantly for the topic of fatigue and safety there have also been pressures towards the reorganization of the truck driver's labour time. An example of such tendencies is contained in five recommendations of the UK National Board for Prices and Incomes in 1967, which were substantially reiterated by the Price Commission in 1976. The recommendations are: (a) a reduction in the length of the working day; (b) the extension of shiftworking (i.e., working round the clock); (c) the adoption of a more flexible working week (i.e., working round the week); (d) increasing speeds of operation; and (e) the use of recording devices (tachographs) to monitor the driver's and the vehicle's perfor-

Table 1.1.   *Numbers of freight employees of major US trucking companies.*

| Year | Company | | | | | | |
	Consolidated Freightways	T.I.M.E.	Denver–Chicago	McLean	Transcon	Roadway	Yellow
1953	3 812	NA	1 680	2 138	502	NA	NA
1959	8 112	1 338	2 518	3 026	1 615	5 221	2 100
1966	12 432	3 079	3 047	5 240	4 444	7 977	2 746
1973	13 628	6 457[a]		9 660	4 520	16 107	11 500

[a] T.I.M.E.–D.C. merger.
Source: Friedman (1978).

mance. These recommendations are designed to increase profitability and productivity in the industry by the increased utilization of capital equipment, and by increasing the intensity of labour.

It is impossible to know the extent to which these changes have been put into practice, but the increased mileage of trucks mentioned earlier is one indication that they have. Another indication is a marked reduction in the length of working time: the average working week of heavy goods vehicle drivers in Britain in 1968 was 58·6 hours; ten years later it was 51·7 hours (National Board for Prices and Incomes 1968, Department of Employment 1979). The amount of shiftworking in haulage companies is also likely to have increased considerably, but there is a lack of statistics to document this. The last 20 or 30 years have seen a large increase in the amount of shiftworking in manufacturing and other industries in most industrialized countries (Carpentier and Cazamian 1977).

However, this should not be taken to imply a uniformity of change within the haulage industry. Many of these changes in the management of the truck driver's labour time can only be exploited to full advantage in large haulage firms with a large volume of regular and predictable business. But as Table 1.2 shows, the numbers of both large and small haulage firms have tended to increase. There is still therefore a large number of small haulage firms. In Ireland, where the development of the haulage industry has been inhibited by a restrictive licensing policy, over 70% of haulage companies have only one vehicle and only 2% have more than six vehicles. In such smaller firms, in order to maintain competitiveness with larger economic units, the pressure must be towards a more extensive (rather than intensive) exploitation of the driver through the prolonging of driving time.

However, as Carré and Hamelin (1978) have shown in their analysis of the French long-distance road haulage industry, the relationship between the large and the small haulage companies is not so much simply one of mutual competition as of dominance and dependence, with the large haulage companies dominating the market, being able to monopolize the most profitable regular operations while subcontracting to the smaller operators the more specialized, difficult or irregular work. The latter are thus forced into increased specialization or into maintaining a rather extreme rhythm and duration of work which enables them to maintain their share of the market. This means that the larger haulage companies are more able to organize their transport operations according to fixed hours with contractual agreements between employer and driver. While it might be thought that these agreements represent the results of resistance of the unionized workforce to excessive exploitation, the fact that these agreements tend to base salary on an average kilometrage and do not fix a maximum, and that they establish norms of work which frequently exceed the regulations, suggests that they serve merely to eliminate only the most extreme hours, rather than improving the norms of work in the industry as a whole. On the contrary, these agreements probably accentuate the problems of subcontractors and independent operators whose workforce is rather more malleable and who are prepared to accept extremes of working hours in order to maintain their independence and to continue

Table 1.2. Public haulage operators analysed by size of fleet: (a) Great Britain and (b) France.

(a) Great Britain

Size of fleet	1963		1979	
	No. (thousands)	% of all operators	No. (thousands)	% of all operators
1 vehicle	23·1	50	67·8	54
2 vehicles	7·8	17	21·6	17
3–5 vehicles	8·2	18	20·0	16
6–10 vehicles	3·8	8	8·4	7
11–20 vehicles	2·2	5	4·4	4
21–50 vehicles	1·0	2	2·4	2
Over 50 vehicles	0·3	1	1·0	1
Total	**46·3**		**125·5**	

Sources: Ministry of Transport 1965, Department of Transport 1980.

(b) France

Size of fleet	1969		1972	
	No. (thousands)	% of all operators	No. (thousands)	% of all operators
1–2 vehicles	20·5	66	17·6	57
3–4 vehicles	4·3	14	4·9	16
5–9 vehicles	3·4	11	4·2	13
10–19 vehicles	1·7	6	2·4	8
20–49 vehicles	0·8	3	1·2	4
Over 49 vehicles	0·3	1	0·5	2
Total	**31·0**		**30·8**	

Source: Hamelin (1975).

to do the kind of work that they like. This situation is more acute in long-distance haulage; in short-distance operations regular hours are more common. In general therefore, there is a division of labour between haulage work that is easy, regular and profitable, and that which is not, and this division of labour depends on the relative control of the haulage market by freight shippers and charterers compared to haulage companies, on the one hand, and on the relationship of dominance and dependence between large and small haulage companies on the other.

1.2. The concept of fatigue

Perhaps the core problem of fatigue research is that of the length of the working day. This was the case in the beginnings of fatigue research in the munitions industry during World War I, and precisely the same question has given rise to much of the current interest in the working conditions of truck drivers. From the point of view of the industrialist the length of the working day crucially affects the output of his employees, in that shorter hours are often more than compensated for by faster work. Hobsbawm (1964, ch. 17) traces the gradual process of realization among nineteenth century industrialists that "low wages and long hours were not necessarily identical with the lowest labour costs", and that "higher labour efficiency implied higher wages and shorter hours" (p. 355). This realization brought the substitution of "intensive" for "extensive" labour utilization and the birth of "scientific management". Fatigue research developed out of this general managerial preoccupation; Vernon (1921) explicitly states the goals of scientific research into industrial fatigue to be to discover for each particular job the optimum hours of work and the optimum speed of work to fulfil the "law of maximum production and minimum exertion".

Hobsbawm (1964) emphasizes the range of factors which determine what is considered 'a fair day's work for a fair day's pay'. The criteria for a fair day's work, he says:

... depended partly on physiological considerations (e.g., the working speed and effort which a man might maintain indefinitely, allowing for rests during and between working days or shifts); on technical ones (e.g., the nature of the jobs he could be expected to do in the course of a day or shift); on social ones (e.g., the need for a team to work at the pace which allowed slower members to keep up and in turn earn a fair day's wage); on moral ones (e.g., the natural pride a man has in doing a job as well as he can); on economic ones (e.g., how much work can earn a 'fair wage'); on historic ones and doubtless on others. (p. 348)

However, fatigue research has traditionally eschewed this broad approach, and adopted a more narrow focus on the measurement of aspects of working efficiency and productivity. These objectives are readily apparent in the concern of the Industrial Health Research Board in Britain, during the first half of this century, in measuring work output and accidents.

Allied to this concern with productivity and accidents was an increasing interest in developing some measurement of the level of functioning of the worker (or experimental subject)—a 'fatigue test' which would allow the condition of the worker to be recorded in a way that was independent of the task that just happened to be being performed at the time.

Muscio (1921) pointed out the methodological pitfalls of such a project, doubting that such a 'fatigue test' was possible, for it would have to fulfil the following two conditions: it would have to be known and stated what was meant by fatigue before an attempt was made to measure it; and different levels of fatigue would have to be measurable independently of any putative fatigue test that was under investigation. Unfortunately these strictures have not always been heeded, and two influential studies which have most blatantly ignored them are those of Jones *et al.* (1941) and Platt (1964), both of which are discussed in more detail in Chapter 4.

Muscio's preconditions for a test of fatigue throw the onus squarely on fatigue researchers to define what they mean by this concept. Perhaps the most elegant and theoretically attractive approach to this problem is contained in Bartley and Chute's (1947) book, *Fatigue and Impairment in Man*. Bartley and Chute make very clear distinctions between subjective experience, physiological state and work output. The concept of fatigue refers solely to the first of these and is defined as follows:

Fatigue is regarded as an experiential pattern arising in a conflict situation in which the general alignment of the individual may be described as aversion. This particular pattern involves feelings of limpness and bodily discomfort which, besides being undesirable in themselves, are frequently taken as tokens of inadequacy for activity. The subjective constituents of this fatigue pattern are not to be taken as epiphenomena, or as symptoms of fatigue, but as fatigue itself. (pp. 47–48)

On the other hand 'impairment' is defined as a "physiologic change in tissue which reduces its ability to participate in the larger aspects of organic functioning"; as such, impairment is never directly experienced, and cannot be deduced from overt behaviour or from bodily feelings. Neither fatigue nor impairment can be measured by the work output of the intact organism. What Bartley and Chute have done is to perform a very neat taxonomic exercise, labelling and drawing distinctions between phenomena. The next stage in the building of a science of fatigue—the delineation of the relationships between these elements—"supposes a science of the person, and this as a fully formed instrument does not as yet actually exist" (p. 401). It still does not, and the fatigue researcher is left with a naive commonsense conceptual scheme of the relationship between feelings of fatigue, some notion of physiological degeneration and a decrement in work performance. It is for this reason that Bartley and Chute's work has had less influence than it deserves on the conduct of fatigue research.

A more operational approach to the problem is represented in the work of Bartlett (1953). He defined fatigue as "a term used to cover all those determinable changes in the expression of an activity which can be traced to the continuing exercise of that activity under its normal operational conditions, and which can be shown to lead either immediately or after delay to deterioration in the expression of that activity, or, more simply, to results within the activity that are not wanted". The majority of studies of fatigue and driving have been founded on a conception of fatigue something like this; thus they have almost exclusively concerned themselves with short-term changes in performance associated with the fact of having driven for a certain length of time. Such phenomena are only a part of what in ordinary language is called fatigue; and reviews by Crawford (1961) and Brown (1967a) emphasized the importance of other factors, such as social stresses and the disturbance of circadian rhythms, in the aetiology of fatigue, which has become increasingly recognized as a very heterogeneous concept turning on a whole range of environmental, task-related and personal factors whose inter-relationships are not amenable to being described in a simple theoretical framework.

A complete re-appraisal of the nature of fatigue, particularly in relation to driving, was produced by Cameron (1973, 1974). Building on a distinction made by Bartley and Chute (1947) between acute short-term fatigue and the chronic syndrome involving cumulative fatigue symptoms, Cameron pointed out that the former pattern, essentially the effect of continued activity on the performance of that activity, is basically the same process as 'reactive inhibition', a process described within Hullian Learning Theory, the effects of which tend to be slight and temporary. The other aspect of fatigue is, in Cameron's words, "embedded in the whole life pattern of those who suffer from it. Periods of days, weeks, perhaps even a whole working life have to be considered" (1974; p. 71). Fatigue in this sense refers to the individual's response to a whole range of stresses to which he is exposed over a period of time. The particular form this response takes—the symptoms of fatigue—is not specific to fatigue. Cameron concludes: "Fatigue is thus a concept which defies precise definition. It is a useful label for a generalized response to stress over a period of time, which has identifiable and measurable characteristics, but it has no explanatory value. It is not legitimate to describe any change in the individual's behaviour as 'due' to fatigue, since the term is no more than a general description of his personal state at the time such changes are noted" (1974; p. 74).

In this book 'fatigue' is used in several senses. First of all it is used to refer to a set of feelings. This set is not closely defined and corresponds to an 'ordinary language' usage of the word, in the sense that it is derived from what people say when they are asked about when they feel fatigue. It will include elements of lack of motivation to continue work or other activity, physical exhaustion, boredom and discomfort. Drowsiness is often distinguished from fatigue in that it relates particularly to feeling close to sleep, though no sharp distinctions are implied. Most contexts in which the word occurs in the literature do not permit a very precise understanding of the nature of the experience involved. If this first sense corresponds roughly to Bartley and Chute's definition, the second is more akin to that of Cameron: 'fatigue' being used as a general descriptor of an open-ended set of phenomena—behavioural, physiological and experiential—associated with a person's exposure to stress over a period of time. There is no presupposition of any logical connections between these different sorts of phenomena, or that different elements in the 'fatigue' complex will always inter-relate in the same way. These are empirical questions. Again, there is no single state called 'fatigue' and its meaning in any one context has to be understood from the circumstances of that context. However, although there may not be any single element common to all states of fatigue, the various phenomena included in it will hold at least a set of 'family resemblances' to each other. Thus the concept does have a certain coherence. In line with this rather tentative and pragmatic usage of 'fatigue', it is enclosed by inverted commas when employed in this sense.

The word fatigue is also used in such phrases as 'fatiguing conditions' and 'fatigue research'. These are no more precise than referring to the sorts of environmental and social conditions that have been associated with 'fatigue' and research into these conditions and their effects.

1.3. *An outline of the contents*

To turn finally to a brief summary of the structure of the book, Chapter 2 is concerned with the social context and physical environment in which the truck driver works, pinpointing a range of stresses to which typically he may be exposed, and their effects on him in terms of his health, his attitudes to his work and his career. Chapter 3 is about accidents—the risks truck drivers themselves face, the dangers to other road users of being involved in an accident involving a truck, and the extent to which these accidents can be attributed to fatigue and adverse working conditions. The three following chapters discuss the experimental evidence concerning 'fatigue' and driving and the implications that may be drawn from it. Chapter 4 concerns driving from the point of view of the skills involved and their susceptibility to degradation. The physiological reactions of the driver are the subject of Chapter 5. Chapter 6 discusses the few studies that have explored the drivers' experiences of fatigue. The seventh and final chapter draws together the lines of evidence that have been developed in the earlier chapters, casts a critical view over the development and current status of 'fatigue' and driving research and discusses some of the practical implications of this research.

Chapter 2
Conditions of Work

2.1. Introduction

In order to understand the relevance of experimental studies concerned with 'fatigue'
and driving, and to interpret the significance of accident statistics, it is important to
know under what sort of conditions the heavy goods vehicle (HGV) driver typically
operates. This involves consideration of such questions as:

1. What statutory limitations are there on drivers' hours of driving, duty time
and rest time? (Section 2.2).

2. What information is there concerning the actual number of hours driven by
HGV drivers; what other duties as well as driving do they typically perform and what
sort of working shifts, rest breaks and off-duty times are in operation? (Section 2.3).

3. Under what sort of physical conditions do they work, particularly in terms of
exposure to factors like noise (Section 2.4), vibration (Section 2.5), extremes of
temperature (Section 2.6), and possible exposure to carbon monoxide gas (Section
2.7)?

4. How do lorry drivers themselves rate the conditions under which they work?
(Section 2.8).

5. Are there aspects of the career pattern of lorry drivers which indicate any
occupational problems? (Section 2.9).

No very comprehensive information exists on all these points; but it is possible to
put together a general picture of the social and working conditions of lorry drivers
from a variety of sources from various countries.

It is not intended to provide a comprehensive list of all possible factors which
could affect the driver's ability to perform safely and well. The factors mentioned

9

above seem to be the most significant potential contributors to 'fatigue' (used here in its most general sense).

2.2. *Legislation*

Most countries, both industrialized and developing, have some mechanism for regulating the hours of driving, work and rest of truck drivers; this is done either through general labour laws, with either specific directives or collective agreements outlining the provisions for road transport, or through specific laws dealing with the hours and conditions of work of drivers. Table 2.1 summarizes the main points of legislation governing drivers' hours in the European Community, the United

Table 2.1. *A summary of drivers' hours legislation in various countries.*

	EEC Regulation 543/69	UK Transport Act 1968[a]	Ireland Road Traffic Act 1961[a]	Code of Federal Regulations Title 49	
				USA	Alaska
Maximum daily driving	8	10	11 (inc. loading, etc.)	10	15
Maximum driving time without a break	4	5·5	5·5	—	—
Minimum break length	0·5 (1 h for heavier trucks with trailers)	0·5	0·5	—	—
Maximum working day/spreadover	12[b]	11		15	20
Minimum daily rest period	11 (8 in some cases)	11	10	8	8
Maximum working week	60[b]	60	—	60	70
Minimum weekly rest period	29	24	—	—	—

[a] Replaced fully by EEC Regulation 543/69 January 1981.
[b] Proposal of the Commission of the European Communities (1976).

Kingdom and Irish acts (which were replaced by the EEC regulations), and the United States regulations including the variations relating specifically to Alaska. There is considerable variation, particularly in permitted hours of driving; and the EEC regulations represent a significant curtailment of driving time in Great Britain and Ireland. It is pertinent to ask the extent to which safety has been an important consideration in framing these regulations.

The legislators themselves have used the argument of improved traffic safety as a reason for the regulations. Thus, the intentions of the hours of regulations of the UK Transport Act 1968 are stated as being of "protecting the public against the risks which arise in cases where the drivers of motor vehicles are suffering from fatigue" (Part VI, section 95, subsection 1 of the Transport Act 1968, quoted in Powell-Smith 1969). Other reasons for the regulation of drivers' hours have also commonly been advanced, in particular the enhancement of the social conditions of life and work of the drivers, and the harmonization of conditions of competition within the industry. The preamble to EEC Regulation 543/69 explicitly refers to the three reasons mentioned above.

There are thus two issues which need to be addressed: What evidence was available to the legislators in framing their regulations? And, were other factors, particularly economic and commercial ones, more important than the question of safety in determining the nature of the regulations?

Those that framed some of the earliest legislation, the 1933 UK Road and Rail Traffic Act, would have had little or no firm scientific evidence to guide them on the question of fatigue and safety, with the exception perhaps of various studies by the Industrial Fatigue Research Board in the munitions and mining industries during and after World War I. They are more likely to have relied on anecdotal evidence (of which much was presented to the Salter Conference on the Transport Industry) and their own judgement to evaluate the causes of fatigue and its contribution to accidents. In the United States, on the other hand, the promulgation of the Interstate Commerce Commission's Hours of Service regulations (which established the 10-hour driving limit) was preceded by two reports by the National Safety Council (1935, 1937). The earlier of these reports concluded that driving excessively long hours was a common practice on American highways, but that total on–duty time (not just driving time), as well as other factors like alcohol and carbon monoxide, were important determinants of fatigue. It also concluded that many motor vehicle accidents occur because drivers fall asleep, or because they are so tired that they are unable to drive safely. The later report described the circumstances of falling asleep accidents: these could occur after any period of driving, and a large proportion involved inadequate sleep during the previous night or nights, a long period of time since the last sleep period, or long periods at the wheel without a break. Slightly later, Jones *et al.* (1941) published their experimental field study which purported to show that various psychological functions related to driving suffered a decrement after prolonged hours of duty, which was taken as evidence in favour of the 10-hour limit which was adopted in the American regulations.

Little further evidence was available by the time of the 1968 Road and Rail Traffic Act in the UK. Indeed, a leaflet issued in 1969 by the Road Research Laboratory concluded that "there is practically no evidence of any correlation between length of driving time and accidents per se or to changes in driving behaviour involving increased risk" (Road Research Laboratory 1969). This leaflet then went on to argue for the 10-hour limit to daily driving periods as a "common sense limitation", invoking arguments primarily concerned with the need to obtain an adequate amount of sleep, and to have adequate rest breaks during the day. It does not seem that the Commission of the European Communities conducted any evaluation of the evidence or commissioned any research pertinent to the problem of fatigue and driving safety prior to the formulation of the EEC regulations.

Thus the empirical support for the formulation of the provisions of existing and past legislation has been rather thin, and only in the case of the United States Interstate Commerce Commission can there be said to have been a coherent attempt to produce and evaluate evidence which would justify the regulations. As far as more recent evidence is concerned, perhaps the most impressive series of studies designed specifically to test the adequacy of a particular set of regulations has been that performed by Human Factors Research, Incorporated for the US Bureau of Motor Carrier Safety (Harris *et al.* 1972, Mackie and Miller 1978, in particular). The BMCS is presumably evaluating this evidence in terms of its implications for amending the US regulations. There is rather less evidence of any dynamic concern with the issue of fatigue and driving safety in the haulage industry on the part of European legislators. An *ad hoc* subcommittee of the Comité de Recherche Medicale of the EEC Commission recently conducted a series of meetings of experts in the field of driving safety but produced no very conclusive recommendations for policy in this area nor a commitment to commission or to undertake any further research on the question. What it did produce was a very general set of guidelines for the direction of future research. O'Hanlon (1979) remarks on the reluctance of the secretariat of the European Conference of Ministers of Transport to discuss or evaluate the EEC regulations in relation to safety, despite widely expressed dissatisfaction in the industry concerning their provisions.

If, therefore, the evidence is rather mixed concerning the diligence with which legislators have sought empirically to justify their pronouncements concerning the impact of their regulations on safety, what of the other expressed reasons behind the regulations?

Considerations relevant to the enhancement of the social conditions of life and work of drivers will of course overlap with the topic of fatigue, which by its very nature is detrimental to the driver's wellbeing, and, when he is driving, potentially threatens his safety. However, they involve rather broader aspects as well, focusing in particular on the discrepancy between the social norms in the pattern of work, rest and leisure within the haulage industry and those in other occupations, which will be elaborated in the next section. Again it is only fairly recently that a systematic investigation of these norms has been undertaken in the haulage industry, and it is impossible to know to what extent this consideration has weighed heavily with the legislators.

It is clear that there are considerable economic and commercial consequences of the regulation of drivers' hours of work and rest. Together with the regulations concerning driving speed and vehicle sizes and weights, they set (or should do) limits on the driver's available labour time and what can be done within it, and hence are one of the basic elements in costs. This suggests that regulatory bodies do have a considerable interest in harmonizing those conditions which affect competition within a market area; and this has been a strong and explicit motive in the promulgation of the EEC regulations (Gwilliam and Mackie 1975). Furthermore, shortening working hours has frequently been proposed as one of a package of measures that will foster increased productivity by promoting an intensification of work and a more efficient utilization of labour time (National Board for Prices and Incomes 1967, Prices Commission 1978).

The issue here is not whether economic or safety considerations have been the real reason behind the regulations, the one excluding the other; for both reasons could be equally valid. Rather it is whether, if the regulations have been framed with economic considerations primarily in mind, they have been framed in a way that does enhance safety and does promote the wellbeing of the driver. The issue is not new, for it has been argued (by Hart in 1959) that the 1933 Road and Rail Traffic Act in the UK made no contribution to safety in the haulage industry, and that the case in favour of the regulations reflected the commercial interests of the larger haulage companies, represented by the Road Haulage Association, against the smaller unorganized haulage companies and owner-drivers. The essence of the case was that excessive competition was making for extreme cost cutting and rate cutting, leading in turn to a high rate of bankruptcies, and that these pressures led to excessive hours of work by truck drivers and poor maintenance of vehicles. The smaller haulage companies in particular were blamed for this supposed vortex of danger.

Whatever happened in an earlier period, a more common contemporary attitude of the haulage industry as a whole has been one of opposition to regulations controlling working practices in the industry (Carré and Hamelin 1978). This opposition has tended to be corporatist in character in that it has been common to employers, large and small, and drivers, both waged and independent owner-drivers. However, these different groups within the industry have different interests in relation to the regulations. Thus the drivers' opinions elicited by Carré and Hamelin in France tended to emphasize the complexity of the regulations and the difficulty of enforcing them; for the employer, the issue of speed restriction was of rather more salience. These authors relate this corporatist attitude to the frequent contact and feeling of a common interest between waged drivers, small hauliers and owner-drivers, and the geographically dispersed nature of the drivers' work, which makes strong organization of drivers difficult, as well as to the ideological role of the aspirations of employee drivers to own and independently operate their own vehicle. However, as this latter option becomes financially and commercially less feasible and less attractive in terms of working conditions, and as it becomes increasingly apparent that opposition to regulations has brought them little or no benefit in their conditions of work, the desire to see these conditions approximate more nearly to the more general industrial norms of working life necessarily involves the acceptance of regulations and a distinction between their own interests and those of their employers, and the development of a greater trade-union consciousness on this issue. Irish drivers whose opinions were elicited by Fuller (1978) and McDonald (1978) variously made comments favourable to the regulations as protecting and improving conditions of work and safety, and expressed worries concerning their economic impact (particularly on their earnings) and concerning their suitability for the industry as a whole.

This brief discussion should surely indicate that the variety and conflicting nature of interests concerning regulations makes it imperative that the issue of safety be treated quite separately from other factors and that it be given a thorough empirical evaluation in all its aspects. The danger is that the issue of safety is used to back up an argument either *pro* or *contra* a particular set of regulations without any such real evaluation of its real role or contribution.

There is a final, crucial, issue concerning legislation, and that is the question of enforcement, for no regulation can be effective if it is not enforced and adhered to. It is not very easy to interpret simple figures totalling the numbers of prosecutions for failure to comply with the regulations in terms of the real effectiveness of enforcement and the extent to which the regulations are flouted. The Commission of the European Communities (e.g., 1980) publishes regular reports concerning the implementation of EEC regulations in the member states. These seem to indicate fairly large discrepancies in the enforcement practices in the different states, and major discrepancies in the amount of information fed back to the Commission on these practices. It is the Commission's view that the harmonization of enforcement policies and penalties and of the reporting of these are prerequisites for the effective implementation of the regulations. There thus seems to be a long way to go before there is an effective Community-wide policy.

The other aspect of enforcement is that of who is liable for the penalties of infringement—the employer, the driver or both. Again, according to the Commission of the European Communities (1980), different rules apply in different member states, with either party, or both being liable. Clearly liability should lie at the level at which control over working conditions and work demands is exercised. The French drivers discussed by Carré and Hamelin (1978) found their liability a considerable problem because, in order to get the job done, and to be in a situation where they can return home for the weekend (and maintain a semblance of a normal family life), it was necessary for them to operate outside the law as a matter of routine. The employer was seen to be in a situation where he can exhort his drivers to respect the regulations while at the same time making demands which (particularly when the inevitable chance delays are taken into account) ensure that the regulations must be broken. The attitude of the state is also seen to be hypocritical in that it is seen to accept and recognize that the regulations are unobservable, and by and large to let things go, but to make spot checks which serve to ensure that the responsibility for the consequences of the actual situation falls squarely on the shoulders of the driver. The driver feels himself to be the victim of a situation over which he has no control. The situation in France may be worse than in some other countries (there is little evidence one way or another) but it is certainly not unique. Rather than being an additional burden for the driver the regulations should be of benefit to him, and should be framed and enforced in such a way as to actually control the norms of work within the industry, and only when the driver, on his own initiative, contravenes such actually existing regulated norms should he be liable for the penalties of infringement.

The significance of this point, concerning the actual norms of work times within the industry, will become apparent in the next section which discusses the hours of work, driving and rest of truck drivers.

2.3. Hours and schedules of work

Detailed information is not generally available from official statistics concerning the distribution of working and driving hours, and the generality of various shift systems. Such statistics do show that in Britain, for example, heavy goods vehicle drivers work a longer week than most other occupations, and that whereas their average weekly earnings may be higher than the mean for all manual occupations (though lower than for all occupations), their average hourly rate of pay is lower and thus a greater proportion of their time is made up by overtime (see Table 2.2). These are average figures and give no clue as to the extent to which very long periods of work and driving do occur in the road haulage industry. The fact that in 1974 in England and Wales there were over 2000 successful prosecutions of drivers for 'failure to observe limits on hours of driving and duty or rest requirements' indicates that the problem does exist, but cannot be considered a reliable guide to its magnitude (see Home Office 1975).

Table 2.2. Average hours and earnings of heavy goods drivers compared to other occupations, Great Britain April 1978.

	Average weekly hours		Average earnings	
	Total	Overtime	Hourly (pence)	Gross weekly (£)
Heavy goods drivers	51·7	10·8	158·1	84·2
All manual occupations	46·0	6·1	172·8	80·7
All occupations	43·1	4·3	204·9	89·1

Source: Department of Employment 1979.

More detailed information is, however, available from a number of surveys of the working conditions of truck drivers which have recently been undertaken in various countries. Thus Hamelin (1975) investigated the operating conditions of 139 lorry drivers in France; his results showed, in his own words, "la médiocrité des conditions de travail dans le secteur du transport routier" (p. 125). Table 2.3 contains his main findings. What seem to be very extreme conditions (certainly in comparison to normal industrial practice) are common, and not surprisingly Hamelin concludes that "oui, les conditions de travail constatées et analysées par nous sont effectivement dangereuses en termes de sécurité routière". Hamelin's sample included short-, medium- and long-distance drivers, as well as international drivers. Generally speaking, the more extreme conditions of long daily and weekly working and driving hours and restricted amounts of sleep are more characteristic of long-distance and international traffic, though still occurring significantly in all types of operation. A number of other surveys in different countries have found very similar results—the pattern of work in the French haulage industry is certainly not unique. Summary results of surveys in FR Germany, the USA and Ireland are presented in Tables 2.4 and 2.5 and Figure 2.1, respectively. In the West German sample there was perhaps a smaller proportion of very long driving hours, suggesting a greater conformity to the regulations (only 1% of drivers exceeded 50 hours of driving per week, the EEC regulations permitting 48 hours). A later American survey (Mackie and Miller 1978) substantially confirmed the distribution found by Harris *et al.* (1972), though

Table 2.3.   *Main conclusions from Hamelin (1975)* Conditions de Travail des Conducteurs Professionnels et Sécurité Routière.

1. The great majority of drivers are married, often with children. Only 42% of the drivers return home every evening; 37% usually being absent all week. 77% of the heavy vehicles they drive have a bunk, which is frequently used by international and long-distance drivers.
2. They drive an average of nearly 10 000 km per month, largely on regular routes. Half of the drivers regularly take part in loading and unloading their vehicles, and a quarter of them at least oversee these operations.
3. The mean length of working week is found to be 62 hours, 30 minutes. 23% of the drivers in the sample work more than 70 hours per week.
4. The mean driving time per week is 38 hours. 27% of the subjects drive 45 hours or more per week.
5. The mean daily working time is 11 hours 30 minutes, but in 34% of days the working time exceeds 12 hours.
6. The mean daily driving time is 7 hours, but 16% of days involve 10 hours or more of driving, and in nearly 7% 12 hours of driving is equalled or exceeded.
7. While the daily distance travelled exceeds 450 km in 33% of cases, long-distance and international drivers drive more than 550 km per day in almost a quarter of days analysed.
8. International and long-distance drivers exhibit a remarkably high frequency of uninterrupted periods of driving exceeding 4 hours—between 11 and 15%.
9. The daily 'spread-over' (the length of the working sequence between two periods of sleep of at least 6 hours) exceeds 15 hours in over a third of spread-overs of the international and long-distance drivers.
10. The mean daily sleep is found to be 7 hours 10 minutes. 31% of days analysed showed less than 6 hours of sleep and 11% less than 4 hours.

Table 2.4. HGV drivers in FR Germany: Weekly hours 'en route' and driving.

Actual driving time (hours per week)	% of drivers	En route time[a] (hours per week)	% of drivers
≤ 10	0	≤ 30	0
11–20	7·2	31–50	12·5
21–30	37·5	51–70	57·9
31–40	34·9	71–90	22·4
41–50	19·4	91–110	4·6
> 50	1·0	> 110	2·6

$N = 304$.
[a] Includes rest breaks.
After Böcher (1975).

Table 2.5. Interstate truck drivers in the USA: Hours of driving and duty on a typical day.

Hours per day	Driving time (% of drivers)	Total duty time (% of drivers)
0–2	7·6	2·2
2–4	13·5	8·3
4–6	16·7	10·9
6–8	19·5	16·7
8–10	37·8	33·8
10–12	3·0	18·2
12–15	1·6	8·5
> 15	0·3	1·4

$N = 370$.
After Harris *et al.* (1972).

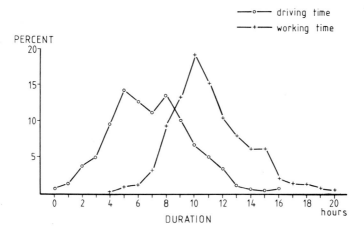

Figure 2.1. Distribution of durations of working and driving hours of Dublin HGV drivers.

suggesting perhaps a greater prevalence of longer times of work and driving: thus 20% of logs inspected showed 15 or more hours of duty during at least one of the previous six days, and one-third of driving times exceeded nine hours; 15% of logs showed duty times between midnight and 0600 hrs (this compared with another American survey which showed 20%). The Irish sample demonstrated that the then current limit of 11 hours per day for driving and associated activities functioned more as a norm for the working day than as a maximum. Unlike most of the other surveys, this sample included a group of local haulage drivers servicing the construction industry whose hours tended to be rather shorter—thus in comparison to the other surveys there may be a sampling bias towards shorter hours in this survey! The hours speak for themselves: the maximum working day in the Irish sample was $20\frac{1}{2}$ hours and nearly 6% of days exceeded 16 hours, one-third exceeded 12 hours, and less than 15% were less than nine hours long. Driving times ranged up to 16 hours and 9% exceeded the 11-hour limit.

Thus, long hours of driving and even longer hours of work are the norm in the haulage industry, and the extremes found must be approaching in many cases the upper limits of the drivers' endurance. As Carré and Hamelin (1978) have pointed out, the working week of the driver is solely dominated by the two factors of working time and time to fulfil the necessary conditions for the regeneration of the drivers' labour power—rest, sleep, food and hygiene. There is no leisure. The focus of all social and domestic activities is thus the weekend (which frequently only stretches between Friday night or Saturday morning and Sunday night when work starts again). Long-distance drivers therefore attach considerable importance to arranging their work to permit them to return home at the weekend and avoid having to spend their weekend in a socially dead industrial zone, guarding their vehicle, waiting for a depot to open. The financial dependence of drivers on long hours of overtime is illustrated by the Irish sample in which basic wages only made up about two-thirds of the total earnings—the balance being made up almost entirely by overtime bonus (except for shiftworkers whose shift bonus made up 10%, overtime bonus 20%). For owner-drivers their work extends even more into their leisure hours with time spent on vehicle maintenance, accounts and commercial relations taking up time that is often not counted or regarded as 'work time'.

The surveys of both Hamelin (1975) and Mackie and Miller (1978) show that long-distance and international drivers are more typified by fewer daily hours of sleep, and a greater proportion of shorter sleep periods, than short- or medium-distance drivers, and this is associated with the fact that the majority of nights spent away from home are spent in the bunks in their trucks. In the latter survey the median sleep period during off-duty periods was over nine hours compared to around six hours in the truck berth, which suggests that the off-duty period is used to catch up on a 'sleep debt' accumulated during the working week. The cab provides an inadequate environment for sleep: insufficient insulation means it can be too hot in summer and too cold in winter; it is noisy if the motor has to be kept running (which is necessary in refrigerated trucks); and if the truck has to be parked near moving traffic there is

additional noise, disturbance from lights, and maybe vibration from air displacement. Not surprisingly sleep in a cab is frequently disturbed (Carré and Hamelin
1978).

There is another important factor which interferes with the ability of many
drivers to get sufficient sleep, and this concerns the time of day during which the sleep
is taken. Shift patterns do have an important influence in determining both the
duration and quality of sleep as has been demonstrated by Forêt and Lantin's (1972)
analysis of the effects of irregular work schedules on sleep. Based on an analysis of
sleep-charts from 40 train drivers who worked very irregular hours their findings
show that:

1. The time of onset of sleep, even when it is interrupted by the beginning of work
practically never occurs before 8 pm.

2. The duration of unbroken sleep periods (either normal nights at home or morning sleep
(after work) is an *inverse function of bedtime* and an almost linear one at that. (p. 279)

The data are given in Figure 2.2. Description of EEG patterns recorded during
sleep confirmed a pattern of 'great rigidity' of the circadian rhythm, with morning

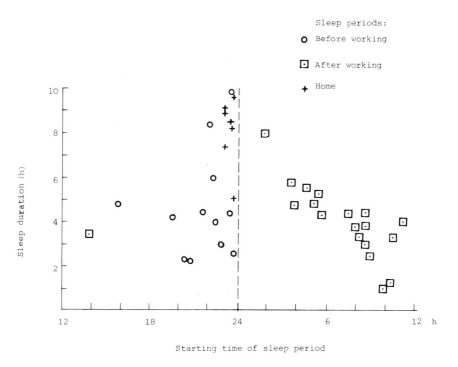

Figure 2.2.   *Relation of recorded sleep duration to the starting hour of sleep.
After Forêt and Lantin (1972).*

sleep showing a much higher percentage of REM sleep than equivalent durations begun in the evening or night. And this is further supported by their finding that all the drivers interviewed found the early hours of the morning—0400–0500 hrs in particular—the most difficult during which to remain vigilant, despite months or years in this kind of job.

Forêt and Lantin's results are mirrored by Mackie and Miller's (1978) study of truck and bus drivers working irregular shifts and 'sleeper operations' (two drivers alternating driving and sleeping en route). Regular day-time relay drivers averaged an hour more sleep than those working a rotating shift; sleep times before a 0300 hrs start averaged five and a half hours and before a 2000 hrs start four hours, compared to an overall average for relay drivers of $6\frac{1}{2}$–7 hours. The 'sleeper' drivers averaged most sleep (around four and a half hours) on trips beginning at 0800 hrs and least (around three hours) on trips beginning at midnight. And reciprocally, critical incidents involving the driver showing symptoms of drowsiness were disproportionately common during the night-time hours.

There is a fairly substantial literature on shiftwork which confirms Forêt and Lantin's findings and provides further evidence of its harmful effects. Shiftwork by its very nature creates a conflict between the requirements of work and the normal physiological, psychological and social periodicities of daily life. Much of the evidence on this has been reviewed by Taylor (1970), Maurice (1975), Rutenfranz *et al.* (1976) and Carpentier and Cazamian (1977). The proceedings of several international conferences also contain a wide coverage of most aspects of the effects of shiftworking on the worker (see for example Swensson 1969, 1972, Colquhoun *et al.* 1975, Rheinberg *et al.* 1981, Johnson *et al.* 1982). Some of the most pertinent findings from this and other research are briefly summarized below.

Aschoff *et al.* (1975) have drawn some general conclusions concerning physiological adaptation to changing sleep–waking patterns: even in 'ideal' circumstances which are conducive to ease of adaptation to a phase shift (as in transmeridian flights, or in conditions of social isolation, where the zeitgebers influencing the circadian periodicity are not in conflict) the process of re-entrainment of these rhythms is slow, varying from a matter of days in the case of the rhythm of body temperature, to a matter of weeks for the complete adaptation of 17-OHCS excretion in urine, for example. Shiftworkers, however, have a sleep–work schedule that is in conflict with the rest of the community, and the evidence suggests that the social zeitgebers are particularly strong, and sufficient to prevent physiological adaptation to the shift. At most there will be a flattening of certain rhythms, less often a partial inversion; and different rhythms may become uncoupled due to the conflicting demands of work and of the worker's family and community (this desynchronization may be particularly marked in rhythms to do with sleep and digestion). Re-adaptation to the normal circadian rhythm is rapid; thus after the weekly days off the process of re-adaptation to the work shift has to begin again. Some authorities recommend fairly long periods of night shift, or continuous night-shift work, as being preferable to fairly rapid alternation between different shifts, thus facilitating adaptation to shiftwork; others recommend very rapid alteration of shifts (every one or two days) which supposedly minimizes any disruption of the normal circadian rhythm. It is clear that the most common shift system—weekly or fortnightly alternating shifts—is the least advantageous (Conroy and Mills 1969). However, adaptation to shiftwork is rarely, if ever, complete and, over the long term, may only be characterized by a slight speeding of the (limited) entrainment of physiological functions to shift changes.

Performance on a range of simple tasks, both laboratory and industrial, has been shown to be worse at night than on the day shift. Typical tasks that have been used in these experiments include measures of response time, signal detection or visual search. Whereas it had been thought that this circadian periodicity in performance was a simple reflection of physiological arousal as manifest in the body temperature rhythm, this view has become increasingly untenable in the light of evidence that, for example, there seem to be different circadian periodicities for different types of task. Thus Folkard (e.g., 1981) has shown that tasks of a cognitive nature (rather than being simply perceptual-motor), which involve a memory component, are better performed at night, but that this circadian periodicity in performance adjusts very rapidly to a changing sleep–waking pattern so that the small relative advantage apparent at night disappears after a few days of night work. It is possible to go some way towards explaining these findings in terms of different optimum levels of arousal for different types of task performance, but this formulation, unless accompanied by independent measures of arousal and task level, can be used to explain anything, but in a non-testable way. A simple unidimensional arousal model cannot explain all the findings, and while Folkard (1981) suggests there may be two or more independently oscillating circadian performance rhythms, Colquhoun (1982) suggests that there is no satisfactory alternative to the arousal model.

Working on the night-shift there is almost inevitably a combination of the influences of circadian performance rhythm, fatigue due to an extended period of work, and sleep deprivation (to at least some degree, depending on the shift system itself and the point in the shift sytem that creates maximum conflict between two sleep–waking rhythms). The effects of prolonged performance at night are to greatly exaggerate the decline in performance efficiency that would have been expected from the normal circadian fluctuation, at least in vigilance tasks (Colquhoun 1982). The effects of sleep deprivation on performance have been explained in terms of an ever increasing number of lapses in performance, with efficient behaviour alternating with faltering responses or no responses at all. Hence self-paced tasks are much less sensitive to performance decrement than tasks in which the pacing is out of the control of the subject. Long monotonous tasks that would lead to lowered involvement and arousal are most affected by sleep loss (Johnson 1982). While these general findings would seem to have strong implications for the driving situation a more extended discussion of prolonged driving and driving at night in terms of performance, physiology and subjective state is contained in later chapters.

Shiftworking severely disrupts the worker's family and social life. Often the shiftworker depends on the wife not working in order to be able to cope with the disruption caused by his working routine. Not only is the general domestic routine affected, but also relationships within the family, marital relations, the role of the father in the education of children, etc. Obviously a lot will depend on particular housing conditions, family size, means of transport and distance to work, among other factors. Shiftworking also adversely affects contact with friends and neighbours (shiftworkers, especially older workers, tend to have fewer friends than day-workers), as well as interfering with a whole range of leisure activities. These and other factors combine to make the shiftworker feel he is not leading a normal life. Most of the studies have shown that the majority of workers are dissatisfied with shiftworking; however there is some ambiguity in the findings. In most studies a substantial minority (occasionally as many as one-half) express positive attitudes towards shiftwork. The main advantages of shiftwork are having more free time and money; the most important disadvantages are seen as relating to sleep, health, fatigue (especially among older workers) and family life.

There is some controversy over the effects of shiftworking on health. The most prevalent complaints are those which concern time-related activities—sleep, digestion and excretion. Both the quantity and quality of sleep are affected, and the sleep of shiftworkers is particularly liable to be disrupted by the noises of traffic, children, and general daily activity. Shiftworkers tend to build up a 'sleep deficit' during the night-shift which they then sleep off during their days off or period of day-shift by extra-long bouts of sleep. Minor problems of digestion and appetite are nearly as common among shiftworkers as sleep problems. Rather more seriously, some reports have suggested that shiftworkers tend to have a high incidence of peptic ulcers, though this has not always been found. Such ulcers could be the result of irregular meals and faulty diet, but may also be psychosomatic in origin due to the stresses of shiftwork. The disruptions caused by shiftwork may lead to a syndrome which has been called 'night worker's neurosis', the symptoms and development of which are described by Carpentier and Cazamian as follows:

General weakness, especially in the morning, insomnia with subsequent sleepiness and character disorders of the aggressive or depressive type. Its development varies. It may appear during the first few months of night work and it may lessen as the worker becomes used to it or require his transfer to a day shift, which will cause it to disappear. In other cases, it appears only after 10 or 20 years of night work, probably in conjunction with the effects of age. In such cases it requires a change of employment, which, however, may not always suffice to make it disappear.

In contrast to these findings there is a large amount of evidence which suggests that shiftworkers have a lower rate of sickness absence than day-workers. The explanation for this is suggested by Aanonsen (1964), who found that while shiftworkers did have a lower rate of medical complaint and sickness absence than day-workers, former shiftworkers tended to have the highest rate. Thus those who find it hard to adapt to shiftwork tend to give it up if possible. Among shiftworkers disorders of sleep, dyspepsia and nervousness tend to occur more often during the night-shift period than the day shift (Bruusgard, 1969). Although no figures are available it is likely that a large number of night-workers take drugs to counteract drowsiness at work and insomnia at home.

There have been attempts to establish some basis for differentiating those who find it easier than others to adapt to shiftwork. Thus, Horne and Ostberg (1976) used a questionnaire to measure the bipolar typology of morningness/eveningness which they related to sleep habits and body temperature rhythm; Folkard *et al.* (1979) added to this dimension the factors of rigidity of sleeping habits and ability to over-come drowsiness; and Colquhoun and Folkard (1978) also suggested that introversion/extroversion and neuroticism (according to the Eysenck Personality Inventory) were related to physiological adjustment, with 'neurotic extroverts' showing better adjustment to circadian phase shifts. However, while it is true that there are considerable individual differences in adjustment to shiftworking, and while factors such as morningness/eveningness are probably associated with adjustment, understanding of the processes involved in adjustment is still limited, and a useful predictive instrument for screening potential shiftworkers is still a long way off. One of the problems is that adjustment to shiftwork involves not only physiological adaptation but the adaptation of the full range of one's nutritional, rest, leisure, social and family needs and demands to the constraints of the 'social time' that is the community norm on the one hand, and those of one's shift system on the other.

Truck drivers are of course susceptible to the effects of shiftwork, as a study by Adum (1975) testifies. Table 2.6 shows his results; both younger and older drivers exhibit the greatest difficulty in adapting to shiftwork.

It is not clear what precisely was involved in Adum's study in the notion of 'adaptation'; however, a number of studies do suggest that there may be occupational health risks associated with truck driving and that the rigorous time schedules drivers fulfil, perhaps associated with the psychological stress and the physical constraints involved in the job, may be implicated in their aetiology. Thus in a comparison of the

Table 2.6. *The difficulty of truck drivers in adapting to shiftwork.*

Age group	No. of drivers	No. of drivers with difficulty in tolerating night-driving	%
26–30	35	7	20
31–35	40	2	5
36–40	17	2	11·8
41–45	17	2	11·8
46–50	20	4	20
50+	16	8	50

After Adum (1975).

health records of a group of truck drivers with groups of bus drivers and air traffic controllers, Gruber (1976) found higher rates of nervous stomach and haemorrhoids in truck drivers compared to bus drivers, higher rates of healed peptic ulcers in truck drivers than in both other groups, and a higher incidence of appendicitis among truck drivers than air traffic controllers (evidence concerning vibration and spinal disorders will be discussed in a later section). Although these three groups were chosen as sedentary groups with different levels of exposure to whole-body vibration, it is clear that while the mechanical effects of vibration may make some contribution to the aetiology of some of these digestive system disorders, other factors may be at least as important, if not more so. Among these, in particular, are the quality and regularity of diet, the disruption of circadian digestive rhythms by working hours, lack of physical exercise, the maintenance of a constant posture for prolonged periods, and the stress of the job of driving. There is some evidence that such factors do form part of the truck driver's working circumstances. Thus, for example, McFarland and Mosely (1954) remarked that long-distance drivers and those working unusual schedules do have very irregular eating patterns and a diet which, while high in protein, is deficient in other respects. Edmondson and Oldman (1974) found a fairly high incidence of overweight and indigestion (as well as backache) in the shiftworking drivers they interviewed; and in both these studies many drivers complained of chronic nervousness, headaches, fatigue, eye strain and exhaustion.

As with other shiftworkers (see Rutenfranz et al. 1977) it is also likely that many drivers have a high consumption of caffeine and nicotine, to counteract the soporific effects of prolonged monotonous driving, particularly at night. These drugs provide a further irritant to the digestive system. Excessive smoking could explain the high mortality ratio for lung cancer among heavy goods vehicle drivers in Britain (Office of Population Censuses and Surveys 1978). In so far as such a smoking habit is engendered by the job the risk of lung cancer must be considered an occupational hazard. Apart from accidents (which will be considered later) the mortality ratios for goods vehicle drivers were not significantly higher than expected, after correction for social class variation, for any other disease category in this population.

Although in Gruber's study bus drivers were compared to truck drivers as a less-stressed sedentary occupation, an earlier study by Gruber and Ziperman (1974) was concerned with the health of bus drivers both longitudinally (in terms of time on the job) and in comparison to other occupations. A range of musculo-skeletal, digestive and circulatory disorders showed a significantly higher incidence among bus drivers of long experience than other groups. These included, in the latter two categories, colitis, gastroenteritis, gastric neurosis, peptic ulcer, appendicitis, diverticulosis, inguinal hernia, haemorrhoids, varicose veins and varicocele. This is rather a broader spread of diseases than those where the truck drivers showed a significantly higher incidence in the later study. This undoubtedly reflects the nature of the comparison groups which in the earlier study would tend to make for larger apparent differences in health status. Once again, however, similar patterns of pathogenic stresses are involved to explain the incidence of these disorders. Valid comparisons on various cardiac disorders and psychiatric problems were not possible in this study because they are subject to a selection distortion in that their presence can disqualify a driver.

This of course illustrates the problem of choosing appropriate comparison groups, and the problem of actually controlling particular occupational characteristics. Thus, to take a small example, the fact that in Gruber's study truck drivers had a lower incidence of nervous fatigue than either air traffic controllers or bus drivers could be part of a pattern in which all three occupations involve some significant degree of stress, rather than an absence of stress in the profession of truck driving, though it is also clear that the nature of the stresses involved in these occupations is rather different. It may well be true that for many bus drivers, whether local or long-distance, the constraints of scheduling are rather more rigorous than is the norm in goods haulage, and certainly bus drivers themselves make complaints of the effects of these schedules on their pattern of diet and digestion, sleep and rest, and family life (Gardell *et al.* 1981, Stress at Work Project Group 1981). However, the epidemiological evidence is not sufficiently fine-grained to allow too subtle distinctions between different driving occupations.

Thus what might be concluded is as follows: truck drivers in the exercise of their profession are frequently exposed to long and irregular hours of work, and/or shift working. There is evidence that professional drivers of trucks or buses may run a higher than normal risk of suffering a range of gastrointestinal and circulatory disorders. The disruption of regular eating patterns, the quality of the diet, and perhaps the stresses involved in the work pattern as well as those involved in the job of driving may make a contribution to some of these disorders, particularly those involving digestive and excretive functions. The evidence from the literature on shiftwork does not indicate a role in the aetiology of serious digestive system disease with the exception of peptic ulcers, where the evidence is contradictory; thus one must be very tentative in making such an inference about professional drivers. Drivers are exposed to a range of stresses in their jobs which may interact in ways that are not necessarily predictable from exposure to each factor on its own; thus it is plausible, though not conclusive, to explain the pattern of morbidity obtained with reference to this combination of stresses. Such an interpretation requires further substantiation however. There is no evidence of an excess mortality for professional goods vehicle drivers for factors other than accidents and lung cancer, and the latter may be explicable in terms of smoking habits.

2.4. *Exposure to noise*

Heavy lorries are a considerable source of noise pollution both for those in the vicinity outside the truck and for the driver and other occupants of the cab. The sources of vehicle noise are the engine, transmission and accessories, road excitation, and air buffeting, which is not very significant below 50 m.p.h. Table 2.7 shows the

Table 2.7.   Origins of commercial vehicle noise.

Origin of noise	Noise inside the vehicle	Noise outside the vehicle
Engine vibration	Major source of low frequency noise	Not important
Engine airborne noise and its transmission	Major source of high frequency noise	Major source of high frequency noise
Engine exhaust	Not important	Major source of low frequency noise
Engine inlet	Not important	Major source of low frequency noise following exhaust
Fan noise	May be noticeable	Can be significant in low and middle frequency ranges
Road-excited vibration	Major source of low frequency noise	Not significant
Road-excited tyre noise	Not significant	Significant

After Priede (1971).

contribution of various major sources to noise inside and outside the vehicle. On the whole, high frequency noise increases with engine speed in a more or less uniform manner, while low frequency noise varies in an irregular manner with vehicle speed, depending on the resonant vibrations of the cab structure. There is, of course, considerable variability in the noise levels within truck cabs. One of the problems is, however, knowing where to measure the noise in order to get an accurate recording of the driver's exposure to it.

In most industrial situations, in broadband and diffuse sound fields, errors in noise measurement due to the location of the microphone on or near the body are small and do not significantly affect exposure measurement. However, in truck cabs the sound field is dominated by discrete frequency components and large measuring errors can result from body reflections and shielding. Substantial discrepancies have been found between the noise exposure dose (expressed as Leq) when measured in the centre of the cab (and according to standard procedures) compared to when recorded from a miniature microphone located in the cavity of the outer ear. Measures taken in a range of truck cabs averaged 5·7 dB higher when recorded from the left ear (nearest the driver's window) than recordings taken from the centre of the cab. Thus errors in exposure dose in excess of 100% seem to be quite possible, with standard procedures underestimating the actual exposure dose of the driver (Reif *et al.* 1980). A range of operating conditions can also have a substantial effect on in-cab noise levels. These include opening the windows, which increases the noise level (by as much as 20 dB(A) is the claim of one report by Kam 1980); using the radio, or CB radio, which can increase instantaneous sound levels by up to 10 dB, and Leq levels by an average of 2·7 dB, an increase in noise dose of around 50%; and variations in the road situation where comparison between city and freeway driving has shown differences averaging 2·9 dB, freeway driving being higher (Reif *et al.* 1980).

Truck cabs also differ in their level of noise. Reif *et al.*'s sample of 58 cabs produced Leqs, measured at the driver's left ear, ranging between 83 and 92 dB(A) in the city and between 87 and 96 dB(A) during freeway driving. When the CB radio was being operated the Leq rose to levels as high as 99 dB(A). These measurements were taken in normal operational conditions in modern cabs (up to 10 years old) which had been well maintained. They are thus likely to underestimate the noise exposure of truck drivers as a whole. These results are generally congruent with other reports of truck cab noise levels in the literature (Close and Clarke 1972, Hutton 1972, Mackie *et al.* 1974, Kam 1980). A discrepant finding is reported by Emme (1970) who reports levels of 100 dB(A) for 80–90% of the time in a number of cabs.

The predominant components of noise spectra of heavy goods vehicles lie in the low-frequency or infrasonic ranges (Priede 1967, 1971, Tempest 1974, Williams and Tempest 1975). Figure 2.3 shows the distribution of noise spectra obtained by Williams and Tempest (1975) for a variety of articulated and rigid goods vehicles. The general pattern shows the highest sound pressure levels to be in the region of 4–16 Hz with intensities of around 100 dB, and falling off to around 80–90 dB at frequencies above 32 Hz. This pattern is reflected in the difference between the average noise level for all the vehicles measured on the 'linear' and 'A' scales: average levels were 107·7 dB(Lin) and 82·6 dB(A).

These levels are in many cases well in excess of the allowable daily 8-hour noise exposure limits framed in regulations and legislation in most industrial countries. In the UK, France and the USA, among other countries, the current limit is a level equivalent to 90 dB(A) for an 8-hour day, though in many countries a level of 85 dB(A) has been set (e.g., Norway, Sweden, Austria and FR Germany). In the USA this latter level is recommended by NIOSH. For longer periods of exposure, such as are common in road haulage (whether actually driving or just being in the cab en route) the allowable limit must be proportionately reduced. Thus, for example, 10 or 12 hours exposure per day is equivalent to adding 0·9 or 1·8 dB respectively to the noise level for the purposes of comparing it to an 8-hour exposure.

However, the major legislative pressure to reduce truck noise emissions comes from specific regulations of vehicle noise limits which have been framed mainly from the point of view of the impact of the noise on the community, though the driver should, of course, also be the beneficiary. The 1980 EEC legislation specifies a noise limit (measured at 7·5 m) of 88 dB(A) for heavy lorries, 89 dB(A) being the current limit in the USA. But the trend in regulation is towards lowering the limits with a goal for ultimate legislation of a limit of 80 dB(A) for heavy trucks. How feasible is this goal?

A thorough discussion of the origins of vehicle noise and the problems involved in reducing or controlling it is contained in a series of papers by Priede (e.g., 1967, 1971,

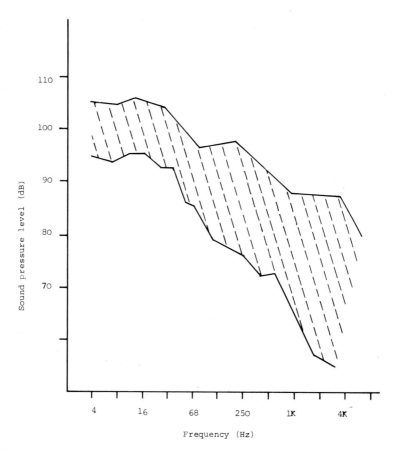

Figure 2.3.   Distribution of noise spectra over a range of vehicles. After Williams and Tempest (1975).

1975, 1979). He concludes his 1979 paper by stating that the control of engine noise by external structural redesign holds promise for some improvement but that the internal structure of the engine is a major limitation in reducing noise. A practical example of an effective truck noise reduction programme is the Transport and Road Research Laboratory's Quiet Heavy Goods Vehicle Project (Tyler 1979) which successfully demonstrated reductions of around 10 dB(A) in two ordinary production vehicles to give external noise readings of 79–80 dB(A) and in-cab measurements of as little as 72 dB(A) in the second vehicle which was modified. This latter vehicle is fully engineered, practical and commercially viable. An earlier demonstration of the effectiveness of various noise prevention measures is graphically illustrated in Figure 2.4 in terms of the percentage of distribution of different noise levels for the various modified cabs. For each modification the allowable driver exposure time is shown,

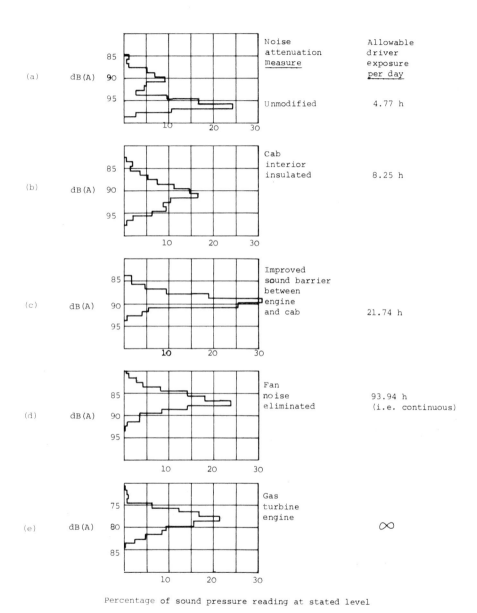

Percentage of sound pressure reading at stated level

Figure 2.4. Summary of noise levels measurements in a variety of cabs. After Hutton (1972).

calculated according to the US Occupational Safety and Health Act of 1970. The quietest condition, a gas turbine engine (note change of noise scale in the figure), has not proved an effective competitor to the internal combustion engine in terms of performance.

Thus it is clear that major improvements in heavy goods vehicle noise levels are technically and commercially feasible, and many of the appropriate measures have been available for some time. The questions are: how long will it take the truck manufacturing industry as a whole to adopt these measures (whether this is a result of pressure from legislators or their customers and the drivers who work in them), when will the 80 dB(A) standard for heavy trucks be adopted in legislation and how long will it take before all excessively noisy trucks currently in production and use are either modified or made obsolete?

What are the likely consequences for truck drivers of working in the noise levels that have been shown to obtain in the industry? Noise has a number of effects which could impair the health, well-being and performance of the driver. Some of these involve the action of noise on hearing and the auditory system, others involve a more generalized psychological and physiological response to noise. Thus the immediate impact of noise concerns the perception and masking of communication, signals, etc. Secondly, high levels of noise affect hearing thresholds both temporarily and permanently and this in turn has a range of consequences both inside and outside the job. Thirdly, the non-auditory effects of noise involve changes in mood and physiological response which have been considered antecedents to changes in performance, general well-being and physical health.

The noise itself can be considered either as a signal or as a masker of signals. Noise from the engine, tyres, etc. does provide an important source of information for the driver concerning the functioning of his vehicle. While such noise may be of advantage to the driver (compared to silence) there is no reason why it should be excessively loud. Other categories of noise might indicate a serious malfunction of the vehicle, such as a tyre blow-out, or some problem, identifiable or not, such as a metallic rattling sound. Other vehicles provide noises such as horns or sirens which it is important to detect. These latter two categories can, of course, be masked by the first. They all indicate the importance of hearing in driving. They also point to a relative balance of advantage in isolating noise sources rather than insulating the driver in his cab, though a combination of both is necessary to protect the driver from unwanted noise. Masking also affects social sounds such as conversation with a companion or listening to the radio, thus preventing the alleviation of the boredom and monotony of long periods of driving. Alternatively, if the masking properties of noise are to be overcome, the loudness of radio or CB radio have to be so loud as to significantly exacerbate the total sound energy within the driver's cab. Sustained attempts to maintain conversation can lead to strain and perhaps damage to the vocal apparatus.

Should drivers wear personal hearing protectors (ear plugs or muffs)? It is clear that, whereas for those with good hearing personal hearing protectors do not impair the detection of pure tones or other warning sounds against a background noise (and may even improve it), this is not the case for those whose hearing is already impaired to some extent. However, on balance the arguments seem to be against hearing protector usage even for those with good hearing, for several reasons: they attenuate noise more at the higher frequencies where most warning sounds have their maximum intensity, they change the intensity and spectral character of the warning signal which may impair recognition or the attentional demand of the signal, and they may alter the psychological state of the wearer leading to a feeling of being more 'cut-off' from the environment (Howell and Martin 1978, Ronayne *et al.* 1982). In support of this conclusion Talamo (1975) reported that tractor drivers were slightly less likely to hear a warning shout against a background noise of 97–98 dB(A) when wearing compared to not wearing earmuffs.

It has long been accepted that there is a relationship between long-term exposure to intense noise and loss of hearing acuity. It is not a simple matter to define optimal exposure limits; as Burns (1974; p. 255) points out: "If all persons reacted in an identical way, a sharp line of demarcation could be drawn which would separate damage from no damage and the problem would then be easily controlled. The variability of normal hearing, the variability of change to higher hearing levels with advancing age, and the variability in susceptibility to noise-induced change makes such an idealistic approach illusory". It is obviously unethical to expose people to potentially damaging levels of noise for experimental purposes, so the major source of evidence for noise-induced hearing loss is from retrospective studies of occupational groups unfortunate to have been exposed to high levels of noise over a long period, correcting for presbyacusis (normal age-related changes in hearing levels) and hearing loss due to disease. Conducting such research involves problems such as having no information on original pre-exposure hearing levels, defining noise levels over the time of exposure, and controlling for the exposure of the subjects to other sources of noise. It is sometimes possible to carry out prospective studies of noise exposure, though ethical considerations prohibit the continuation of the noise exposure beyond the first evidence of a change in hearing level.

The first sign of hearing loss is a depression in the audiogram between 3 and 6 kHz, commonly at 4 kHz, and is not subjectively noticeable. This depression deepens with continued exposure, but the rate of deterioration broadens to higher and lower frequencies, depending on the spectral characteristics of the noise. Although temporary threshold shifts due to acute noise exposure occur in particular frequencies according to the frequency bands of the noise, evidence from industrial studies seems to show the dip at around 4 kHz to be the predominant feature of chronic exposure to extremes of both rising and falling noise spectra (Burns and Robinson 1970). On this basis it has been possible to demonstrate a mathematical relationship between hearing levels and noise exposure (expressed as noise immission level—N.I.L.) (Robinson 1970, Robinson and Cook 1970). N.I.L. is roughly the amount of sound energy reaching the ear, weighted according the 'A' scale of noise intensity, and is a function of both intensity and duration of noise. Thus one can calculate the proportion of people likely to suffer any level of hearing loss for any duration and intensity of exposure. However, the range of effects is very large; there is a greater scatter of obtained hearing loss with duration of exposure, and considerable hearing loss can arise without any pathological basis in ears not exposed to excessive noise (Burns 1973).

It is quite likely that damage-risk criteria based on the 'A' weighting under-estimate the damaging effects of noise with strongly falling spectra (Burns and Robinson, 1970); such spectra are typical of heavy goods vehicles (see Figure 2.2). Furthermore, it has been suggested that the combination of noise and vibration (such as occurs in trucks) makes for a higher temporary threshold shift in hearing than exposure to noise alone; and the increase in temporary threshold shift is accelerated at high temperatures (Okada 1972, Manninen 1983 a, b).

The evaluation of hearing loss in terms of 'impairment' is also a subject of controversy (Kryter 1970, Burns 1973). Impairment is usually taken to involve social disability, the most logical criterion of which would be in terms of deterioration of speech perception. However, indices of hearing impairment have been based on pure tone thresholds. The most widely accepted index is that recommended by the Committee on Conservation of Hearing of the American Academy of Ophthal-mology and Otolaryngology (AAOO 1959), which is the average of hearing levels at 0·5, 1 and 2 kHz, with a hearing level of 25 dB being taken as the cut-off point for mild handicap. More recently the National Institute for Occupational Safety and Health (NIOSH 1972) has recommended taking the mean of the hearing levels at 1, 2 and 3 kHz, on the grounds that hearing loss at 3 kHz is more predictive of difficulty in speech discrimination than a loss at 0·5 kHz.

It would be expected on the basis of both the typical levels of noise found in truck cabs and the evidence from other occupational groups concerning the effects of chronic exposure to intense noise, that many truck drivers would suffer noise-induced hearing loss. This has been confirmed by Mackie *et al.* (1974), who measured the hearing thresholds of 45 drivers by pure tone audiometry, and found higher thresholds at many frequencies than would be expected on the basis of presbyacusis (see Figure 2.5). According to the AAOO criterion only two of the 45 drivers would be defined as having some degree of hearing impairment; but on the NIOSH index eight of the 45 drivers (18%) would be considered hearing impaired. Nearly all the drivers in this sample had more than 18 years of driving experience (with a median of 26 years). Mackie *et al.* unfortunately had no data on possible previous exposure of these drivers to other high noise environments. In a similar study Brohm and Zlamal (1962) found that all of 51 drivers in their sample, having been exposed to truck noise levels of 90–110 dB(A) for periods of 10–30 years, had audiograms showing moderate to severe hearing loss.

The consequences of this loss of hearing are perhaps most profound outside the work situation and can involve increasing social isolation and a lower level of participation in social and recreational activities. Domestic and family relationships can also be adversely affected with stress and tension caused by failures of

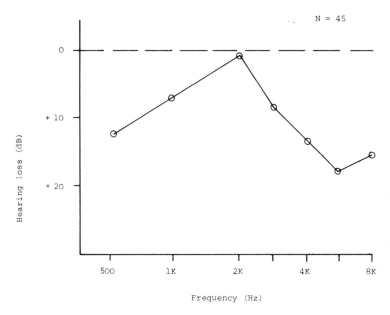

Figure 2.5. Median hearing loss in a group of truck drivers, corrected for presbyacusis. After Mackie et al. (1974).

communication and the partially deaf person's need to have television, radio, etc. set at a very loud volume. This difficulty of communication frequently makes for a general process of social withdrawal (Ramsdell 1961, Sataloff 1966, Myklebust 1964, Lundahl 1971). Similar social difficulties will tend to occur at work where, for example, messages may be misunderstood. Conversely, in the cab, the loss of hearing acuity perhaps makes the noise more tolerable as its perceived loudness decreases; however, by the same token sensitivity to auditory signals and sounds will be impaired. It is worth mentioning also that the effects of noise on hearing levels is first manifest in a temporary threshold shift which persists for up to a few hours; while there has been little investigation of the effects of this threshold shift, it is likely to involve increased difficulty in understanding conversation, particularly against noisy household or social backgrounds, with the consequent frustration that this entails.

Physiological response to exposure to occupational noise is characterized by an increase in physiological activation, and is known as an 'arousal' or 'stress' response. This generalized response, which is conceived of as mobilizing the body's resources for action, involves a range of bodily systems—central nervous, neuroendocrine, cardiovascular, gastrointestinal, etc. In addition, under certain circumstances there may be reactions specific to the noise stimulus. Thus, there are reports of desynchronization of the EEG, and of increased secretion of catecholamines and corticosteroids in the blood and urine (though in the latter cases this has not been consistent, and individual differences, particularly in anxiety, may be important). The cardiovascular system responds with peripheral vasoconstriction, and while there are reports of increases in blood pressure, the typical cardiac response in humans to peripheral vasoconstriction is a compensatory decrease in stroke volume, though those with established hypertension do show an increase in hypertensive response. There is also evidence of cardiac abnormalities (e.g., dysrhythmias, T-wave depression) which may be due to coronary vasoconstriction. Changes in the respiratory rhythm, decreases in digestive secretion and gastric motility have also been reported. Visual and vestibular aberrations have also been reported, but only at levels of noise unlikely to be encountered by the truck driver. This area is reviewed by McLean and Tarnopolski (1977), Schiff (1973) and Ronayne et al. (1981).

Many of the studies demonstrating these reactions have been laboratory studies, in which one might expect a heightened 'stress reaction' due to the novelty and unusual nature of the situation, and the question arises as to whether these changes represent a more or less temporary 'orientation response' which will habituate on repeated exposure to the noise or whether there are any long-term physiological consequences which could result from chronic exposure to intense noise. The evidence is not absolutely clear-cut on this question. While for some people in some circumstances there does seem to be habituation of the physiological response (Glass and Singer 1972), in other situations this does not occur (Jansen 1969). However, the fact that many components of this pattern of physiological response have been demonstrated in people exposed to noise at work would seem to suggest that habituation does not occur. The argument partly hinges on whether the noise has continuing meaning or significance to the person exposed, which would prevent habituation. Thus the role of annoyance in reaction to noise might, for example, be a crucial factor in maintaining a chronic stress response, though Jansen would claim this response to be independent of annoyance.

There is controversy as to the possible long-term consequences for health of exposure to industrial noise. The most convincing evidence suggests an increased incidence of hypertension in noisy industries (Anticaglia and Cohen 1970, Parvizpor 1978, Johnsson and Hansson 1977). However this has not always been found (Drettner et al. 1975, Lees and Roberts 1979). One of the problems in epidemiological questions of this kind is that exposure to noise is likely to be confounded with a range of other potentially pathogenic stresses such as danger, dirt, heat, shiftwork and other pressures of work. It is impossible to definitively separate out the different contributions of these factors. It is also difficult to take account of individual differences in susceptibility to a potentially noxious agent like noise. Thus it is not clear whether noise has a role in the aetiology of hypertension, or whether it acts solely to exacerbate an already existing condition. There is no evidence from the studies of Gruber and Ziperman (1974) or Gruber (1978) that bus or truck drivers suffer an increased incidence of hypertension or other cardiovascular problems. It is quite possible that those who are susceptible or show an early onset of symptoms are rapidly selected out of the industry as a result of mandatory periodic medical examinations.

Among the most immediate psychological reactions to noise are feelings of annoyance and discomfort. Industrial studies have shown that as the noise level increases so does the number of workers reporting annoyance or discomfort (Öhrström et al. 1979, Manninen 1977). A wider range of emotional response has also been reported in a laboratory study where subjects working in noise reported increased feelings of anxiety and depression, less social affection and lowered concentration (Jones and Broadbent 1979). Other laboratory studies have suggested that people tend to be less helpful to others in high levels of noise and show a greater tendency to be aggressive when provoked to anger (Mathews and Canon 1975, Geen and O'Neal 1969, Donnerstein and Wilson 1976). While the situations investigated in these experiments were not directly comparable to the truck driver at work, the possible implications of these findings are clear.

The complaints of workers who are chronically exposed to industrial noise tend to emphasize feelings of tiredness and fatigue and a range of minor psychosomatic and neurotic symptoms including headaches, anxiety, disturbances of sleep, irritability and a tendency to have more social conflicts at home and at work (Bugard et al. 1953, Jansen 1961, 1969, Dumkina 1973, Öhrström et al. 1979). It has been suggested that annoyance due to noise can give rise to an impoverishment of mental health, and indeed a small but significant relationship has been found between the two variables McDonald et al. in preparation). But it is possible to argue that causality operates in the opposite direction in that certain people's mental disposition might predispose them to be sensitive to noise. And it should be noted that a direct relationship between noise exposure and mental health problems has not always been found (Lees et al. 1980).

A variable which has been shown to crucially affect whether or not the effects of noise occur in terms of physiological response, mood and task performance, is whether or not the persons exposed to the noise consider they have any control over the noise source. Perceiving that one has control over the noise, whether or not this control is exercised, has been shown to mitigate the adverse consequences of noise (Glass and Singer 1972, Sherrod et al. 1977, Lundberg and Frankenhauser 1978). However, the relevance of this to driving a noisy truck for a living is not great. Whereas in one sense the driver is in control of the machine which creates the noise, in fact he has little control at all over his noise exposure.

There is a fairly long history of research into the effects of noise on the performance of various types of task, though few of these studies have been undertaken in actual industrial situations; they are, rather, laboratory simulations of certain industrial tasks.

One of the few industrial studies was undertaken in a film processing laboratory. The particular experimental manipulation involved the reduction of noise from 99 to 89 dB in one room and comparing it to another room where the noise was unchanged. The rate of work increased in both rooms relative to other parts of the factory, probably due to an improvement in morale (an instance of the 'Hawthorne effect') but there was also a significant reduction of film breakages in the quieter room

(Broadbent and Little 1960). Among the other industrial studies that have reported a detrimental effect of noise are those by Kovrigin and Mikheyev (1965), who found that the number of sorting errors made by postal sorters increased systematically with noise level, and deAlmeida (1950) also reported that absenteeism was reduced in an electric punch card room when noise levels were reduced.

Broadbent and Little interpreted their results in terms of the hypothesis that the effect of noise is to increase the frequency of momentary lapses of attention and co-ordination—in effect one is distracted by the noise from the task in hand. This hypothesis had earlier been investigated in the laboratory using a variety of different sorts of task. Typically these tasks involved monitoring and responding quickly and accurately to a display of dials and lights, representing (presumably) the functioning of complex process control machinery. A number of studies with such tasks have found that performance is worse in higher rather than lower levels of noise, particularly in more difficult tasks, and that performance in noisy conditions increasingly deteriorates as time goes on (e.g., Broadbent 1950, 1951, 1954, Broadbent and Gregory 1965).

Performance during high levels of noise is also likely to be more variable than in quieter conditions. Again this is predictable from the theory that noise causes distraction and lapses in attention.

Obviously if noise acts in this way it will tend to affect some tasks to a greater extent than others. Tasks which are externally paced and which do not allow the performer to make up for lapses of attention or 'periods of distraction' will be particularly vulnerable to disruption (Broadbent 1953, Wilkinson 1963).

On the other hand a number of studies have found that noise is associated with an improvement in performance. These tasks tend to be relatively undemanding on the subject either in terms of the inherent complexity of the task or the subject's control over pacing (Davies 1968, Cohen 1973).

There are also qualitative differences in performance in high levels of noise in that some aspects of task might be affected more than others. High levels of noise have been found to affect the selectivity of attention; the performer will focus his or her attention on more central aspects of a task or combination of tasks, to the detriment of those aspects which are less important or in the periphery. Typical experimental investigations on this pattern have used two simultaneous tasks, one designated as the primary task and the other as the secondary. Performance on the primary task typically is maintained at a high level or even improved during noise whereas performance on the secondary task suffers (Boggs and Simon 1968, Hockey 1970, 1973).

Thus, in general, summarizing the results of a large number of studies it can be said that noise will adversely affect tasks which are complex rather than simple, and paced rather than unpaced. Typically the effect of noise is to increase the number of errors made, to produce faster decisions, more of which are incorrect, and to increase the variability of performance. There is also an increasing narrowing of attention towards those aspects of the task which are perceived as being more central and important.

The explanation of the effects of noise has progressed beyond the distraction theory to include the concept of arousal. This was necessary in order to explain the fact that noise can often improve performance on simple tasks, and to account for complex phenomena like the narrowing of attention. This increase in arousal is presumed to underlie the narrowing of attention and the blocking of irrelevant information that accompanies noise.

This theory is not, however, without its critics; and Poulton (1976, 1978 a) has sought to explain many of the detrimental effects of noise on performance in terms of the masking of auditory ones in the laboratory equipment and the masking of 'inner speech', which presumably controls performance. Broadbent (1978) has sought to rebut these criticisms.

What implications have these findings for the skill of driving? While it is clear that the basic skill of keeping the vehicle moving along the road and avoiding obstacles is simple and overlearnt and not easily subject to disruption by noise, there are certain aspects of the task where this may not be the case, for example, in complex decisions in interacting with other vehicles and following a route, and maintaining vigilance for an unexpected event which may well arise at the periphery of the central task. The driver may also be particularly vulnerable when driving under strong time pressure, which reduces his control over pacing. A more detailed analysis of the skill of driving and its vulnerability to degradation is the subject of Chapter 4.

One of the major predictions of the arousal explanation is that for any particular task there is an optimal level of arousal that makes for the best performance of that task. Too low or too high a level of arousal will be associated with a lower level of performance. Thus, simple, tedious work is performed best, and fastest, with a high level of arousal and motivation, whereas more complex work is best done feeling fairly relaxed. Noise, by increasing the level of arousal, can thus improve performance on tasks where the arousal level is too low (most likely on simple tasks); and make for a deterioration in performance when the arousal level is too high (most likely on complex tasks).

This has suggested to some people that an arousing agent like noise can be used to counteract the detrimental effect of working at night, or working after an accumulated loss of sleep (which is also associated with low arousal). The noise will maintain the operator's alertness and improve his performance (Poulton 1978 b). The logic of this argument is that in the interests of safety one should maintain a high level of noise and other discomforts in truck cabs of shiftworking drivers in order to keep them awake and maintain their attention on the road ahead.

This effect (improving the level of performance of operators who are suffering from a loss of sleep by having them work in high levels of noise) has in fact been demonstrated in the laboratory (Corcoran 1962).

However there is one major flaw in the argument and this is that such effects have only been demonstrated over a very short period of time—about $\frac{1}{2}$–1 hour. It seems highly unlikely that such an effect could be sustained for a whole working day or life. This is because prolonged physiological activation or arousal must have some cumulative physiological and psychological cost. This cost is shown in the feelings of fatigue and disturbances of mood that have been associated with exposure to noise, and which were discussed earlier.

The arousal hypothesis can also be invoked to explain individual differences in the effects of noise in findings like those of Cohen *et al.* (1966) that the performance of anxious subjects was impaired more by high noise levels than was that of non-anxious subjects.

Some investigators have suggested that low frequency noise and infrasound at levels which frequently occur in motor vehicles can have adverse effects on the driver (Rao 1975), including the induction of "a false sense of well-being or euphoria" which could be a contributory factor in single vehicle accidents (Tempest 1972). However it would seen that these reports are rather exaggerated, and the laboratory studies which have purported to show an adverse effect of infrasound at intensity levels comparable to those likely to be found in road vehicles have been severely criticized on methodological grounds.

Hearing thresholds for infrasonic frequencies are considerably higher than for the 'audible' frequencies (as the terminology implies). Thresholds are frequency dependent and Yeowart (1972) found binaural thresholds to vary between over 120 dB at 2 Hz to over 85 dB at 20 Hz. Effects that have been reported at fairly low above-threshold intensities within this frequency range include vertical nystagmus, sensations of swaying, and performance decrements in reaction time and pointer following tasks (Evans 1972, Evans and Tempest 1972, Hood *et al.* 1972). In a thorough and critical review of these and other studies Harris *et al.* (1976) suggest that the poor reporting of experimental designs, statistical tests and significance levels makes it impossible to substantiate the performance findings, that there is no proper evidence for a nystagmus effect and that reports of swaying sensations could well be a result of suggestion. There is thus no convincing evidence of an adverse effect of infrasound at levels which would affect truck drivers. Under more extreme intensities characteristic of military jets (up to around 150 dB between 2 and 50 Hz), chest-wall vibrations, gag sensations and respiratory–rhythm changes have been observed, and at higher frequencies (up to 100 Hz) headaches, choking, visual blurring and post-exposure fatigue have been reported after very brief exposure (Mohr *et al.* 1965). However, these are not conditions that occur in trucks.

Thus, in summary, the noise exposure of truck drivers is likely to be underestimated by the standard noise measurement procedures. Although noise levels vary according to operational conditions, and the use of the radio, it is clear that under normal operational conditions a significant proportion of trucks expose drivers to levels of noise in excess of a Leq of 90 dB(A), and many more are over 85 dB(A). And it should be noted that the driver's exposure is usually greater than 8 hours per day. This situation persists even though the technology for controlling the noise entering the cab has been available and described in the literature for some time. On the positive side, the legislative trend is towards specifying lower noise levels, and increasingly manufacturers are producing quieter trucks. However, the problems of noisy trucks will persist at least until all noisy vehicles currently in use and still being produced are either modified or end their working lives.

Most importantly, these levels of noise will continue to be a threat to the driver's hearing in a situation where personal hearing protection is inappropriate. They will also continue to suffer from a range of other consequences of noise exposure which may affect their efficiency and safety of performance (masking of sounds and signals, possibility of impaired performance under certain circumstances) and their psychological well-being. Prolonged exposure to noise is also likely to contribute to fatigue, and this non-specific effect of noise may in turn contribute to a deterioration of the driver's performance.

2.5. Vibration

As well as being noisy, truck cabs also vibrate, and the 'ride' of the cab is an important aspect of the driver's environment. "The forces impressed upon the driver can be in many directions and have a variety of pathways—viz. the seat pan, the seat back, the foot controls and the steering wheel. The vibration sources are assumed to be the engine vibration and the response of the vehicle to surface irregularities. The vibration levels can be quite high especially if the truck is unloaded" (Sandover 1975).

Unfortunately it is not easy to specify at all precisely what might be the effects of this stress on the driver. This is largely because of the complex way in which vibration impinges upon the body. The problems are neatly summarized by Schoenberger and Harris (1971): "quantification of vibration responses will never reach the accuracy achieved in acoustics, due to such factors as the lack of a unique receptor for vibration, the multiplicity of vibration transmission paths, and the fact that vibration transmission to the body in the resonance range may be greatly altered by changes in body position and muscle tone". The bulk of research has involved vertical vibration, particularly sinusoidal, and has been carried out in the laboratory. Less is known about response to horizontal vibration, vibration in different directions at the same time, or with a combination of different frequencies.

Nevertheless it is possible to draw some general conclusions which are applicable to most situations. The greatest sensitivity of the human body to vertical vibration is between 1 and 20 Hz, with the maximum transmissibility at around 5 Hz (Coermann 1961). Above 20 Hz the vibratory motion is increasingly attenuated by soft tissues and the effects more localized to the point of contact. The major resonance in the horizontal axes is in the region of 2 Hz; the effects of horizontal vibration depend very much on the way it is applied to the subject (Guignard and King 1972). Subjective assessment of vertical vibration also shows a greater sensitivity at around 5 Hz; however, this is qualified by the considerable variability of results from a number of different studies. The tremendous range of thresholds obtained for perception, discomfort and intolerability is remarkable (for perception the highest threshold is

100 times the intensity of the lowest); also the overlap between the various zones, to the extent where vibration conditions found by some experimenters to be below the level of perception approach what other investigators have found to be above the level of tolerance (Hanes 1970).

Guignard and King (1972), notwithstanding the variability of the above results, summarize the intensity-related effects of vibration as follows: "Laboratory studies of the limits of human voluntary tolerance have indicated that as a rough rule for z-axis whole-body vibration exposure, sinusoidal vibration in the most critical frequency band (4–8 Hz) is likely to be physically uncomfortable at acceleration amplitudes much above 0·1 g; painful or distressing at intensitites in the region of 1 g; and injurious at acceleration amplitudes exceeding 2 g if sustained for more than a few cycles of motion" (p. 67). Typically, physiological response to whole-body vibration is not unlike the response to moderate exercise. Heart rate increases, particularly at the onset of vibration, blood pressure increases and vascular resistance decreases. Respiration volume increases (though probably not respiration rate) and there is increased oxygen consumption, associated with the increased metabolic cost of the muscular work needed to control posture, and a reflex increase in muscle tension (particularly at 10–20 Hz). Increases in body sway and difficulty in controlling posture and movement have also been reported. Hyperventilation can take place under strong whole-body vibration at low frequencies, due to mechanical driving of the respiratory system and/or excessive simulation of certain proprioreceptors which causes inappropriate reflex increases in respiration. Prolonged or severe hyperventilation can cause hypocapnia, but this is more a problem in military aircraft than heavy goods vehicles. Endocrinological effects are concomitant with a non-specific stress response (Guignard and King 1972).

Recognizing the potential adverse effects of exposure to whole-body vibration, the International Standards Organization (ISO 1978) has proposed recommended limits for exposure to whole-body vibration, though in interpreting these limits in practical situations the paucity of comprehensive and unequivocal data must be born in mind. Three criteria give three sets of limits: the 'preservation of working efficiency' criterion giving the fatigue decreased proficiency (FDP) boundary; the 'preservation of health and safety' giving the 'exposure limit'; and the 'preservation of comfort' criterion giving the reduced comfort boundary. Von Gierke, the chairman of the committee which produced the standard, describing the philosophy behind and the development of the standard, has emphasized the following points: the standard covers as many experimental data and practical situations as possible (hence there is a tendency to simplify and generalize); and it is conservative (i.e., it tends to err on the side of safety). The frequency response curves are "reasonable envelopes to the experimentally observed curves of constant human response", notwithstanding variability due to posture, support, etc. The time dependence curves were based on a few studies concerning physiological response and subjective judgement of comfort and performance efficiency, and the logical requirement that the relationship between the comfort, FDP and exposure limits be preserved. Again, it is emphasized that the standard is not to be seen as "setting firm limits but (as) a general guide for the evaluation of vibration exposure" (von Gierke 1975).

There have been criticisms of the standard; indeed von Gierke's (1975) paper was a response to an attack on the standard by Janeway (1975). Janeway recommends the concept of 'constant absorbed power' as being a more theoretically elegant basis for a standard than the more pragmatic generalizations of the ISO committee. This concept was developed by Pradco and Lee (1968) and refers to the rate of energy absorbed by the subject's body, and is equivalent to the concept of a noise 'dose' in acoustics. However, as von Gierke points out, the fact that the human body can be seen as comprising a number of separate mechanical systems with different resonance frequencies, these again varying with posture, means that human response cannot be a simple function of absorbed power.

In a more sympathetic and balanced review, Guignard and King (1972) suggest that these recommendations represent the best available summary of research evidence, and that tentative guidance concerning exposure to vibration is better than none; however, they do provide a critique of the recommendations, emphasizing the following points: they do not take account of individual differences of response, or situational factors affecting response; they assume independence of response to each frequency and to each axis of vibration; they do not take account of work rest-schedules or interactions with other stresses; and the relationship between the three recommended sets of limits (proficiency, comfort and health) is not an empirical one, but arbitrary. Floyd and Sandover (1972) similarly criticize the ISO guide-lines, emphasizing that terms like 'fatigue-decreased-proficiency' are too vague; that the guidelines need to be stated in terms of confidence limits—what proportion of such a population is likely to be affected; and that what is needed in vibration research is a greater emphasis on observation in field situations, categorization of exposed populations and development of task taxonomy, and the development of a more standarized experimental technique, the lack of which has been a major factor in the variability and inconsistency in findings from vibration research.

The problem is, nevertheless, one of evaluating the impact of vibration on the comfort, performance and health of truck drivers. First it is necessary to review evidence on the levels of vibration in trucks and some procedures for isolating the driver from the sources of these vibrations.

Sources of vibrational input to the vehicle include road irregularities, tyre irregularities (imbalance or run-out), and at higher frequencies vibration from the engine, transmission, etc. These inputs are highly modified by the resonance properties of the vehicle structure (tyres, suspension, chassis, frame, fifth-wheel coupling between tractor and trailer, cab structure, etc.). Thus, for example, Jex *et al.* (1981) showed that while peaks of power spectral densities (recorded from the floor of a tractor-trailer cab) had frequencies which varied directly with speed, and were related to wheel asymmetries and periodic roadway inputs, the underlying shape of the power spectral densities was the same at all speeds. At the truck's 'resonance speed' these peaks coincided with the dominant vibration frequency (about 3·6 Hz). Some vibration measures (vertical vibration at the floor) declined with increasing speed above this resonance speed. Comparing ten different trucks over five different Californian road types, a significant influence of road type on vibration levels was demonstrated, though this was particularly marked for accelerations measured at the floor, less so for those measured at the seat, where higher frequency vibrations tended to be filtered out (and in some trucks vertical accelerations around 3 Hz tended to be amplified). A much more extreme example of variation between road types is given by Ribarits *et al.* (1978) for a Volvo, comparing European and Middle-Eastern roads, making the difference between just exceeding and massively exceeding the ISO 8-hour FDP boundary (though these levels were measured at the cab structure rather than at the seat). Jex *et al.*'s cabs were generally within the ISO 4-hour FDP limit for the best trucks, and within the 1-hour limit for the worst trucks. They did not find it possible to account for the variation between trucks in terms of particular design and load features as the sample was too small to encompass the many possible combinations, and the difference between these factors could have been overridden by maintenance differences. Overall, then, the picture emerges of some variability between trucks, but of there being considerable room for improvement, even in countries with a highly developed road infrastructure; and further, the situation as outlined by Jex *et al.* does not seem to have changed in the almost 10 years since Stikeleather *et al.* published their findings (see Figure 2.6).

In terms of reducing the driver's exposure to whole-body vibration the important facts are that wheel and terrain induced acceleration levels tend to be high in the region of maximal human sensitivity, and that truck frame natural resonances also tend to be in this frequency region. Furthermore, in cab-over-engine cabs the driver is in a very unfavourable position in relation to the main vibration nodes and is therefore subject to considerable pitch accelerations, also known as 'cab nod', due to chassis beam vibrations in long-wheel-based vehicles and vehicle pitching in short-wheel-based articulated tractors. The mechanical solution to the problem of isolating the driver from this vibratory input favours very soft suspension systems with natural frequencies well below those that are causing the problem. The laws of transmissibility through a sprung mass system dictate that when the ratio of forcing frequency to natural frequency is one, resonance occurs; at frequency ratios less than one, transmissibility is greater than 100% for all forcing frequencies; and at frequency ratios greater than one, the transmissibility is greater than one until a frequency ratio of $\sqrt{2}$ is reached; greater than this, transmissibility is less than 100% and declines as the forcing frequency increases. Hence the further below the forcing frequency the natural frequency is, the better.

Efforts to reduce the vibration exposure of truck drivers involve either the reduction of the natural frequency of the chassis suspension or isolating the cab or the seat with a suspension system of low natural frequency. One of the problems with soft suspension systems is that they involve large deflections. In suspension seats this means high amplitudes of driver movement in relation to the cab and controls. Also suspension seats do not control pitch. With cab suspension systems there are problems in controlling cab movements such as roll 'brake dive' and 'squat'. Control of pitch suggests that the cab should optimally be suspended at driver height, but this is impractical. For chassis suspensions, the huge difference between empty and loaded weights of trucks is better managed with stiffer tyres and springs with higher natural frequencies. For cab and chassis suspensions the greater clearance required in soft suspension systems has cost disadvantages (Foster 1978, Jex *et al.* 1981). Reducing the pitch frequency of a truck tractor to 2 Hz or less is claimed to be the most effective method of reducing the vibratory input to the driver (Herider and LeFevre 1967). However, suspension seats are by far the most widespread solution, though cab suspension systems are becoming increasingly common.

The effectiveness of suspension seats in reducing the vibration input to the driver is demonstrated in Figure 2.6 (*a*). Foster (1978) suggests that air suspension seats are best, having the lowest natural frequency. In contrast, a foam cushion on a non-suspended seat resonates at around 3 Hz and amplifies the input to the driver at this most sensitive region (Figure 2.6 (*b*)). Some suspension seats also incorporate fore and aft vibration isolation. Ribarits *et al.* (1978) describe two cab suspension systems, one with an articulated and the other with a rigid truck. They conclude that cab suspension was able to solve the problem of frame-dependent vibration in the rigid truck, but did not substantially improve acceleration levels in the articulated truck in the critical frequency range. On the other hand, Foster (1978) claims considerable reductions in absorbed power with a Ford cab suspension system; however, these reductions are again most marked in the higher frequency ranges, above the critical sensitivity region.

Some validation of these efforts to improve suspension systems is provided by Miller's (1981) finding that the engineering design feature which made the clearest difference to overall evaluation of ride quality by drivers was cab suspension (of the same type as described by Foster).

From the foregoing, therefore, one can conclude that, certainly in terms of the ISO norms, there is a vibration problem in most trucks in current normal operating circumstances, and in some trucks this is a serious problem. Substantial improvements are possible by suspending seats or cabs and by improving axle suspension. But just how serious is this problem in terms of the driver's performance, comfort and health? It is necessary to go beyond the ISO guidelines and interpret research findings in terms of the driver's task and situation. There are two broad issues which need to be elucidated: what is the mechanism through which vibration has its effects, and how does this process change over time?

Direct mechanical effects of vibration will impair performance on tasks requiring a certain degree of fine visual acuity and manual control. Concerning visual acuity, the effects of vibration will depend on whether the observer or the visual object or both are being vibrated, and on the posture and support of the body (which will determine how the various frequencies and axes of vibration are transmitted to the ocular mechanisms). When the observer is moving, the vestibulo-ocular reflex

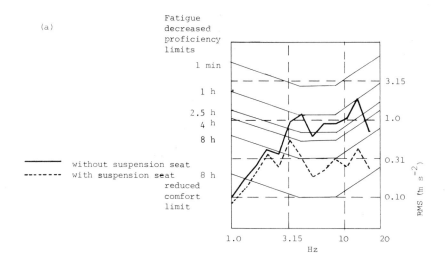

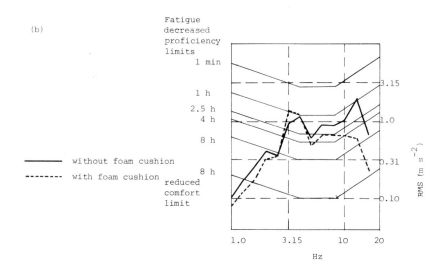

Figure 2.6.   Vertical vibration spectra for various vehicles compared to ISO fatigue decreased proficiency and reduced comfort limits: (a) for a tractor-trailer unit (ride simulator) with and without a suspension seat; (b) for a tractor-trailer unit (ride simulator) with and without a foam cushion; (c) for a tractor-trailer unit (empty) on an expressway (55 m.p.h.) at the man/seat interface; and (d) for a compact car on an expressway (60 m.p.h.) at the man/seat interface. After Stikeleather et al. (1972).

(c)

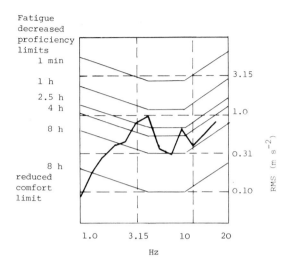

(d)

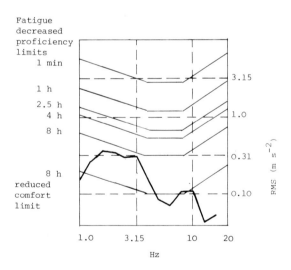

enables the ocular muscles to compensate for this movement and the eye to track the target. These compensatory movements are much more efficient than the pursuit movements required to track a moving target. They may be beneficial below about 5 or 10 Hz; at higher frequencies vision is possible by fixation of the nodal points of the oscillating image. There are, however, great individual differences in the frequency of maximum sensitivity to the effects of vibration, and some subjects have more than one frequency of maximum sensitivity. Griffin and Lewis (1978), in their comprehensive review of this area, report a frequency-dependent curve of the level at which approximately 50% of a group of subjects reported some degree of visual impairment (see Figure 2.7); comparing this with Figure 2.6 and the results of Jex *et al.* would seem to indicate that a significant proportion of trucks would induce at least some visual impairment in a proportion of drivers. However, in this study the subjects were asked to adopt postures which resulted in the maximum sensation of vibration in the head, and the authors conclude that adopting postures for least vibration transmission would only occasionally have resulted in blurring.

As far as the road and traffic environment is concerned, it is the driver who is in vibratory motion in relation to the visual object; however, the driver's eyes are also in motion in relation to the moving cab structure (and this includes the instruments and controls) because of the distortions and damping introduced by the seat structure and the driver's body. In this situation reflex compensatory eye movements will be

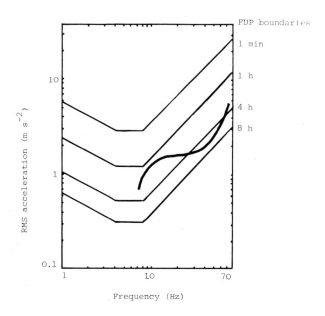

Figure 2.7.   *Levels of vertical whole-body vibration above which reductions in visual acuity may occur, compared with levels given in ISO 2631. After Griffin and Lewis (1978).*

destructive of visual acuity, particularly at low vibration frequencies around 1–3 Hz. The significance of this in relation to the fairly gross visual discriminations required within the cab during driving is maybe not grave, but it does underline the importance of exceptionally clear and unambiguous design of instruments and controls. More seriously, vision through cab-mounted mirrors is likely to be impaired.

Griffin and Lewis (1978) were unable to find any data to suggest time-dependent effects of vibration on visual acuity, though such effects might result from changes in posture over time (for example). There is a lack of evidence concerning the effects of vibration (time-dependent or otherwise) on other visual functions, such as visual search, peripheral colour vision, or those dependent on attention. Such functions may be more significant in the driving task than acuity *per se* (see Chapter 4).

In a companion paper to the above, Lewis and Griffin (1978) review the literature on vibration and manual performance. In terms of driving trucks, their most significant conclusions are the following: the largest increases in error in tracking tasks occur when the vibratory motion is in the same direction as the sensitive axes of the control and display (this is not critically the case in driving); at frequencies of less than 20 Hz the decrements in performance are related to the transmission of the vibration through the body (rather than through its effects on neuromuscular or sensory-motor processes), hence for sinusoidal vibration the biggest decrements are in the 4–5 Hz region in the vertical axis, 1–3 Hz in the horizontal axes. Less is known about the effects of complex vibration spectra and of different axes of vibration operating together. While most of the studies reviewed concern tracking with a joystick, others have found that tracking with a steering wheel is affected by vibration in the horizontal and vertical axes, at levels between 0·15 and 0·35 g. Brake reaction time is not affected (Schmitz and Simons 1959, Hornick 1962, Chenchenna and Krisanankutty 1969).

These effects of vibration are direct mechanical effects and as such are unlikely to change over time; indeed there is little or no evidence of a decrement in performance of these laboratory tasks. An exception is Hornick's (1962) study; however, the decline in performance accuracy that he found during $1\frac{1}{2}$-hour's vibration exposure is more probably attributable to the effects of task performance than to the vibration. Overall, this appears to be directly contrary to the predictions of the ISO guidelines, and has caused difficulty for some commentators to reconcile this discrepancy with either what seems a common-sense view, or with the lack of any reasonable alternative to the ISO time-dependent curves. However, von Gierke (1975; p. 4) has made it absolutely clear that: "the recommended exposure times are for daily routine occupational (habitual) exposures for extended periods, even a lifetime", and he questions the relevance of once-off laboratory experiments in this question. What he is proposing is therefore a completely different mechanism for time-dependent vibration effects, one which is based on the role of vibration in producing fatigue and undermining the motivation to perform well. It is interesting that the only non-laboratory study concerning prolonged exposure to vibration quoted by Lewis and

Griffin (1978) demonstrates precisely the point von Gierke is making: Jackson (1956) analysed the records of altitude and heading of aircraft during a series of 15-hour flights. During the first half of the flights the pilots tended to fly more accurately and consistently in rough air than in calm air. In the latter part of the flights, however, they were adversely affected by turbulent conditions; thus suggesting some interaction between fatigue and vibration.

The notion of vibration as a non-specific stressor is not unknown in laboratory studies, particularly where vibration is combined with other stresses such as noise or heat, and is hypothesized to act as an arousing agent. Again, a number of these have been reviewed by Lewis and Griffin, who conclude that the interaction of vibration with other stressors is complex and cannot be accounted for simply by arousal theory. Perhaps, also, a clue to the solution of the conundrum is provided by a small finding by Grether *et al.* (1971) who found that on performance measures (tracking and choice reaction time) vibration on its own had the greatest detrimental effects, less when combined with heat or noise or both; on the other hand subjective ratings of stress increased with the number of stressors. This would seem to indicate that performance in such situations does not accurately reflect the experience of stress, being strongly affected by the novelty, artificiality, and social demands of the experimental situation, and it therefore cannot be taken to represent performance in the 'real world' under conditions of prolonged experience of stress.

One commentator who has enthusiastically adopted the logic of experimental studies of vibration as an arousing agent and applied it to the everyday situation of truck drivers is Poulton (1978 c), who seems to lament the passing of uncomfortable cabs, which, he argues, keep drivers awake and alert. There is something fascinating about the naive empiricist logic of this argument which utterly disregards the special circumstances of the experimental situation. One can only imagine the response of a professional truck driver to Poulton's suggestions.

Thus, in summary, direct mechanical interference by vibration on sensory and motor control processes has been demonstrated. The significance of these findings for the truck driver's safety is not absolutely clear, and the topic of performance and safety will be discussed more fully in Chapter 4. Vibration has also occasionally been shown to improve performance in laboratory tasks, presumably due to an arousing effect, but direct comparison of this effect with the driver's everyday life task is unwarranted. A time-dependent decrement in performance due to vibration has rarely been demonstrated. This situation reflects a lack of understanding of the different mechanisms whereby vibration might affect performance, and a failure to study realistically those processes which may be affected by prolonged exposure to a stressful situation in normal working life.

It is a major disappointment that there is so little evidence concerning the effects of exposure to whole-body vibration on fatigue and task performance in normal occupational settings. Jex *et al.* (1981) found in their survey of truck ride that the drivers' ratings of ride discomfort were not so severe as might have been predicted from the ISO comfort and FDP boundaries and suggest that these might be doubled (in effect this would leave the worst trucks in their sample below the 4-hour boundary, and the best trucks below the 8-hour limit). However, firm conclusions should not be based on such short periods of driver exposure, and where lack of familiarity with the test vehicle and immediate comparisons with the vehicle of habitual use might to some extent influence judgements. Further comments on the role of vibration in driver comfort are contained in Section 8 of this chapter.

On the question of health, the strongest evidence implicating vibration in trucks with some impairment in health concerns back pain and spinal damage. Although there is a history of clinical reports associating an incidence of spinal disorders with truck, bus or tractor driving, which dates back to 1950 (Fishbein and Salter 1950), there is as yet no conclusive evidence upon which to base an exposure limit like that proposed in the ISO guidelines. The most recent evidence implicating the driving of trucks in the incidence of spinal disorders comes from epidemiological studies by Kelsey and Hardy (1975), Gruber (1976) and Frymoyer *et al.* (1980). The largest sample of truck drivers (over 3000) was contained in Gruber's study, in which they were compared with groups of bus drivers and air traffic controllers (sedentary occupations exposed to respectively decreasing levels of whole body vibration). Truck drivers had a higher incidence of both vertebrogenic pain syndrome and premature degenerative deformations of the spinal column than either bus drivers or air traffic controllers; however, the difference between truck drivers and air traffic controllers in the case of vertebrogenic pain syndrome was not statistically significant, which seems surprising, and, as Troup (1978) points out, Gruber seems to have misclassified the spinal deformities in the ICDA codes, throwing some doubt on this result. Kelsey and Hardy's sample were 217 patients with acute herniated lumbar intervertebral discs, matched with a group of other patients and an unmatched control group. They found that truck drivers were nearly five times as likely to develop this condition as matched controls, and were over twice as likely as the unmatched controls. Those who had never been truck drivers also had a higher likelihood of the disease. These associations were independent of any lifting activity. Both sedentary occupations and driving (non-professionally) were also associated with significantly higher likelihood ratios of herniated lumbar disc, though not to nearly the same extent as driving trucks. Frymoyer *et al.,* studying the incidence of low back pain in the patient population of a family practice, also found a significant association between the incidence of low back pain and truck driving. These are all retrospective studies which are subject to the distortions due to non-standardized measurements of diagnostic procedures, and to selection biases within occupational groups. However, together they provide good prima facie evidence of an association between driving trucks and back trouble. One can hope that a more comprehensive picture emerges from the extensive study proposed by Naughton and Pepler (1981) for the National Highway Traffic Safety Administration in the United States. The question remains as to the role of vibration in the aetiology of spinal disorders.

Troup (1978) points out that back pain can only arise from those tissues which carry a nerve supply, and hence that an intervertebral disc can be damaged, and minor fractures of the vertebrae sustained, without pain or disability. However, if repeated, these injuries have a cumulative effect leading to an early onset of degeneration. He identifies four sources of spinal stress which could result in either back pain or in diagnosable spinal damage: postural stress, vibration, muscular effort, and impact and shock.

The normal effect of gravity on the spine is a loss of height during the day (with recovery at night); this is due to the loss of fluid from the disc which becomes less compliant, and results in an increase in the area of contact between the bearing surfaces of the joints of the spine. This process, known as creep, is accelerated by additional load on the spine, and also by vibration ('vibrocreep'). This will change the dynamic response characteristics of the spine when exposed to shock or impact.

Secondly, the isotonic muscular activity required to maintain a given posture over a given period may induce muscular fatigue. When the spine is subjected to postural stress for long enough it stiffens as well as shortens. The neuromuscular control of spinal posture and of reactions to external forces is likely to be modified.

Thirdly, when seated the lumbar spine is flexed, and there is greater intervertebral pressure on the discs than when lying, standing or walking; when seated, driving a car, the disc pressure is increased when the gear is changed, or when the foot pedals are depressed (Nachemson 1976). The muscular effort required to stabilize the body when braking, accelerating and cornering and in response to bumps or lurches of the vehicle will also increase pressure on the discs.

Fourthly, exposure to repeated shock and impact may increase the susceptibility of the spine to injury, particularly when it has been conditioned by the sum total of static, vibratory and muscular stress throughout a driving shift. The physiological response to minimize impact or shock is to tense the muscles, but if one cannot anticipate the shocks (as in driving) this is impossible. Although most of the data on spinal fracture when seated involve injuries caused by ejection from aircraft, there is the possibility that shock and impact in trucks may cause fatigue failure of the vertebral bone and perhaps the disc. It is also likely that the incidence of fracture is higher than suspected.

While it is quite clear that the postural, vibrational and muscular stress involved in driving a truck may cause backache, and that this will be more common and disabling in those who already have a lumbar spine disorder, there is little direct evidence that these stresses (as opposed to impact or shock) can actually cause spinal damage. However, Kelsey and Hardy's findings concerning sitting and driving seem to indicate that these postural and muscular stresses do have a role in the aetiology of herniated lumbar disc.

In contrast to this it has been suggested that because levels of vibration imposed on the body during walking or jogging are in excess of the ISO exposure guidelines, these guidelines cannot possibly represent unsafe levels, as walking or jogging are 'natural' activities (Barton and Hefner 1976). Quite apart from the question of the 'naturalness' of continuous walking or jogging for an 8- or 10-hour day (or longer), the biodynamic considerations outlined above would seem to vitiate any meaningful comparison between walking or jogging and sitting in a moving truck or heavy machine. On the contrary, reducing vibration levels will reduce spinal stress in the driver. Hence the ISO guidelines, notwithstanding the lack of precise evidence for their detailed formulation, are much better than no guidelines. Reducing vibration by improved suspension will also attenuate the shock or impact from major road irregularities. The question of shock or impact has not received the attention it deserves considering its potentially serious consequences for the spine. There is no published evidence concerning the incidence and severity of shock in truck cabs. It is also possible that some weighting factor for shock or impact should eventually be incorporated within exposure guidelines for vibration, perhaps in an analogous way to the practice in acoustics where impulse noise is increasingly recognized as having particularly severe effects.

Adequate seat design is also crucial in protecting the driver's spine. The following considerations are important: The seat cushion and back should be adjustable to allow an incline of 10° or more for the cushion and 15° or more for the back. There should be good lumbar support which should be adjustable to suit all drivers (or exceptional drivers should have their own individually moulded spinal support). The seat back should also provide good lateral support. The height of the seat should be adjustable, and the front edge rounded, so that the weight of the thighs is not borne by the underside of the thighs, but where there is gentle support for the thigh muscles, and no abrupt changes in pressure distribution. Other ergonomic improvements to increase reduced muscular work and back strain would include power steering, and ensuring that the line of action of pedal force should pass from the foot through the hip joint (Herider and LeFevre 1967, Troup 1979). It is regrettable but true that these basic ergonomic principles of seat design are not universally held in the industry.

2.6. Temperature and ventilation

"The provision of heating and ventilation equipment in trucks has been a comparatively recent development in climates such as that of the U.K. . . . apart from the comfort aspect, a well designed system provides the driver with an environment in which he can remain alert and will therefore be safer than the man who is trying unsuccessfully to keep warm, or alternately is fighting off drowsiness arising from an inadequate ventilating system" (White 1973). There are no statutory standards for heating and ventilating equipment in truck cabs.

Man, being a homeotherm, maintains a core temperature within fairly narrow limits, usually between 36·4 and 37·2°C. There is some slight diurnal variation, and some variation between the different sites at which core temperature is measured (the mouth and the rectum are the most common sites) (Leithead and Lind 1964). It seems likely that it is hypothalamic (or internal cranial) temperature which is regulated (Carlson and Hseih 1965), but oral and rectal temperatures are the most adequate practical measures. If core temperature is lowered to around 32 or 33°C the physiological temperature regulation mechanism begins to break down; shivering is replaced by permanent muscular rigidity, and there is a gradual loss of consciousness. Death normally occurs before body temperature reaches 25°C (Carlson and Hseih 1965, Poulton 1970). At the other end of the scale, when core temperature is elevated to around 39 or 39·5°C some people would be likely to suffer acute heat disorders, be unable to continue work and may collapse. Sweating, the main physiological cooling mechanism, may fail at around 40·6°C; and death usually occurs when core temperatures reach 42–43·5°C. The range of core temperatures broadly compatible with functional efficiency is 36–39·5°C (Leithead and Lind 1964).

Intensity of exposure to adverse temperatures is not a function of air temperature alone. Air speed, radiant temperature and relative humidity are also important, the last particularly in hot environments where it affects the rate of sweat evaporation. The radiant environment in trucks includes heat radiation from the engine as well as solar radiation.

In very cold temperatures, clothing is the most effective method of heat insulation (Burton and Edholm 1955) and the severity of exposure will depend on the effectiveness of the clothing. The main physiological response to cold is shivering, which increases metabolism and produces heat. It is not a very efficient method of maintaining body temperature, as much of the insulation of the tissues is reduced by increased circulation, and there is probably also increased convection loss due to body motion (Carlson and Hseih 1965). These authors also mention another method of heat production, called 'non-shivering thermogenesis', which is related to the release of catecholamines. Skin temperature is normally cooler than core temperature, and decreases in skin temperature will increase the depth of cooling. Temperatures in the hands and feet show the greatest effect of cooling and this is related to decreases in blood flow (produced both by peripheral vasoconstriction and the greater viscosity of cool blood). Acclimatized people show a less extreme effect (Burton and Edholm 1955, Carlson and Hseih 1965). Lowered temperatures in the extremities are accompanied by numbness and reduced sensation, as well as by a decline in hand strength and dexterity. Performance decrements in a wide range of manipulative tasks have been reported (Poulton 1970). Lockhart (1966, 1968) independently varied hand and body skin temperatures and found that decreases in each of them adversely affected the subjects' manipulative ability, which suggests that the decrements in manual dexterity may involve more central processes. Other evidence that CNS processes could be affected by small changes in body temperature is provided by Poulton et al. (1965), who found that lookouts on the bridge of a ship performed a vigilance task worse in the Arctic than in a more temperate climate. Average oral temperatures of the sailors in the Arctic fell by 0·7°C to 35·6°C during the watch. In so far as cold temperatures could adversely affect driving, Provins (1958) suggests that the reduction of tactile discrimination and other forms of kinaesthetic sensitivity could reduce the driver's 'feel' of the vehicle, particularly over icy roads, especially at night, or in other conditions where visibility is reduced.

Many factors affect the intensity of, and susceptibility to, heat stress. These range from climatic factors such as air temperature, humidity, radiant temperature and air movement, to situational factors such as the amount and type of clothing worn and the rate of work, to personal factors such as body weight, and possibly age. There are two main physiological mechanisms for reducing body heat: the cardiovascular system responds with increases in pulse rate and peripheral vasodilation which raises blood flow and transfers heat from the core to the periphery; the dissipation of heat from the skin is then accomplished by the process of radiation and convection and by the evaporation of sweat, and the latter is by far the most effective of the two response modes (Leithead and Lind 1964, Fox 1965). Acclimatization is characterized by a progressive reduction in rectal temperature and pulse rate, following the increases found on first exposure; the amount of sweat increases and it is more dilute; skin temperature decreases; cardiac output declines to pre-exposure levels; and peripheral blood flow, though declining after the rise on initial exposure, remains at a higher level than in a cool environment. An increase in circulating anti-diuretic hormone (ADH) reduces the amount of water excreted through the kidneys, counteracting the loss due to sweating. Drinking sufficient liquids to offset the sweat loss prevents dehydration and makes for faster acclimatization, though this is not always possible under high levels of heat stress. Increased salt intake during the first few days of exposure is also advisable. Acclimatization is faster in humans who are in good physical condition (Leithead and Lind 1964).

There have been a number of attempts to construct a heat stress index which will weigh accurately the more important factors which determine heat strain. Leithead and Lind (1964), Fox (1965) and NIOSH (1973) provide comprehensive reviews of these. None of the indices are free from criticism, but five which have received most consideration are the Effective Temperature (ET) scales, the Wet Bulb Globe Temperature (WBGT) index, the Predicted Four-hour Sweat Rate (P4SR) index, the Belding–Hatch index, and Fanger's (1967) Basic Comfort Equation.

The ET scales are based on empirical investigation of subjective impressions of warmth in any combination of wet- and dry-bulb temperatures. The corrected version (CET) substitutes globe temperature for dry bulb to take account of radiant heat. There are different scales for different amounts of clothing, but different rates of work are not catered for. These scales are most accurate as an index of comfort in warm temperatures where activity is light; but in higher levels of heat stress or during heavier work they are not very reliable. The WBGT index shows similar scale values to the CET scale, and is probably subject to the same criticisms; however, it has the considerable merit of simplicity. It weights the relative contribution of globe (g), dry bulb (d) and wet-bulb (wb) temperatures in the following way:

$$WBGT = 0.2\,g + 0.1\,d + 0.7\,wb$$

The next two indices, the P4SR and the Belding–Hatch, are both based on estimates of sweat production as the criterion of heat stress. There is considerable evidence that sweat production is a good measure of physiological strain experienced in response to heat stress, though individuals vary a lot in the amount of sweat produced and neither index can be used to predict an individual's sweat rate. They are indices of climatic stress and not of physiological strain (Leithead and Lind 1964). The P4SR index encompasses a wide range of combinations of climatic factors (wet-, dry-bulb and globe temperatures and air movement) as well as levels of clothing and rates of work. Lind concludes that it is the most accurate index of heat stress, except under extreme conditions where the upper limit is reached to the amount of sweat which can be produced (Leithead and Lind 1964). The Belding–Hatch index is an attempt to quantify the total heat-exchange process between the body and its environment. Heat stress is expressed as the ratio between the amount of sweat that has to be evaporated to maintain thermal equilibrium, to the maximum evaporative capacity of the environment. It takes into account a wide range of factors, including height and weight of the subject, and skin and rectal temperatures. However, it does make a number of assumptions for the sake of simplicity and has not proved as accurate as the P4SR in predicting the physiological effect of different climates.

Fanger's (1967) Basic Comfort Equation is based on the close association of thermal comfort with both skin temperature and sweat secretion. Two separate equations are derived to describe sweat secretion and skin temperature at thermal comfort as a function of activity level. Optimal thermal comfort levels can be calculated for different activities, clothing, and levels of the climatic variables (temperature, humidity and air speed). The combined equation shows good agreement with independent empirical results with sedentary subjects; it is less accurate at higher levels of activity. Fanger's equation is probably the most complete index of thermal comfort available.

While it is not possible to give a complete account of possible exposure to heat stress in an environment such as that of the truck driver, Mackie *et al.* (1974; p. 21) summarize the main considerations thus:

Physiological signs of heat stress are evident in most humans at WBGT's above about 79°F (26·1°C). The earliest signs, elevated heart rate, increased skin blood flow and sweating are associated with the heat adaptation process. That process should be capable of maintaining thermal equilibrium until the WBGT exceeds about 86°F (30°C), when the body begins to store heat. Body temperature as determined from rectal and oral thermometer readings will rise beyond that point. In the case of truck drivers, whose expected rates of metabolic heat production are in the range 2–3 kcal/min . . . , body temperatures should stabilize at levels which increase by about 0·25°C (0·45°F) with each additional 1·0°C (1·8°F) increase in WBGT.

Mackie and his associates consider that typical climatic conditions in many parts of the USA during summer, where temperatures are over 80°F (27°C) WBGT for much of the time, represent a real danger to the truck driver.

Recommended exposure limits of the American Conference of Government Industrial Hygienists are based on the maintenance of body temperature below 38°C. Thus for continuous light work the threshold limit value is 30°C WBGT (ACGIH

1975). However, it seems likely that there is an upper thermal tolerance limit for efficient performance of various types of task and that this limit is well below the physiological limit which is based on possible adverse health consequences. Wing (1965) extrapolates such a limit for exposure durations of up to 4 hours from a number of experimental studies of different types of mental performance. Performance decrements have been found in a wide range of tasks in hot conditions; these include such intellectual tasks as electrical circuit-testing, coding and decision-making, as well as tasks requiring considerable physical effort and motor skill. Some industrial studies have found reduced production and higher rates of minor accidents to accompany higher temperatures, though as with many industrial studies it is hard to attribute this unequivocally to heat (Poulton 1970). Subjects in the studies on which Wing based his 'exposure limit' were all unacclimatized to the conditions under which they were performing. Acclimatization to heat increases the optimum temperature for task performance and the range of temperatures at which people are most comfortable; exceeding these temperatures, will, of course, lead to discomfort and a falling off in efficiency.

To ensure the driver's maximum comfort and efficiency, the heating and ventilation system should be able to maintain his whole cab at around 25°C (according to Fanger's equation), which is the generally preferred level for seated people. Relative humidity is not quite so important for the driver as for someone doing more active physical work; though many drivers do suffer from the effects of perspiration due to prolonged sitting on non-absorbent seats and this is likely to be exacerbated by high humidity. High radiant temperatures from the sun are also likely to be a major source of discomfort. The driver's performance is likely to deteriorate within a few degrees above his preferred temperature and he may become irritable or drowsy. Cold is easier to counteract, by clothing; however, the driver should not have to wear bulky and constricting clothes in the cab, and he has little opportunity to produce heat through movement. Adequate heat to all parts of his body is therefore essential. If he has to perform tasks out of doors like loading and unloading, or roping and sheeting, then he is potentially exposed to extremes of temperature; however, there is no reason why his cab should not provide an adequate thermal environment.

2.7. *Exposure to carbon monoxide*

Carbon monoxide as a possible danger to the driver could arise from three sources: from general levels of carbon monoxide in the air in cities or rural roads with particularly heavy or congested traffic, from a direct leak of CO from the exhaust system into the cab, or from smoking cigarettes. McFarland and Moore (1957; p. 891) summarize the effects of CO thus: "Very small amounts of this gas will be rapidly absorbed by the blood stream, resulting in an oxygen deficiency that may at first be unnoticed by the driver. The initial reaction to carbon monoxide consists primarily of lowered attention, difficulty of concentration, poorer night vision, slight muscular inco-ordination, sleepiness and a mental and physical lethargy. These first symptoms are not permanently injurious but, owing to their nature, could easily involve the driver in hazardous situations."

Horvath *et al.* (1971) found that the state of hypoxic stress produced by levels of CO found on well travelled roadways impaired their subjects' ability to sustain attention over a monotonous vigilance task. McFarland *et al.* (1944) and McFarland (1970) have found that levels of CO in the blood concomitant with heavy smoking have significant detrimental effects on the rate of dark adaptation, and on visual acuity in the dark. However, McFarland (1973), while finding some evidence that relatively high blood levels of carboxyhaemoglobin (11% and 17%) may possibly affect peripheral vision and the amount of attention required to control a car at speed, found no effect of lower levels (6%), more typical of those produced by smoking, or by the air in cities, on either visual, complex psychomotor, or driving tasks. Some of Rockwell and Weir's (1973) results also suggest that COHb (carboxyhaemoglobin) levels of between 6 and 14% may affect some visual aspects of driving: subjects with elevated COHb levels showed greater 'perceptual uncertainty' (their glances tended to be longer, and the time they were prepared to have their vision obscured was shorter); also a slight increase in average speed and a reduction in mean following distance were associated with elevated COHb, which could possibly reflect a deterioration in speed perception (Section 4.2 includes a discussion of the perception of speed). Ramsey (1970) found that simple reaction times of drivers were longer after inhaling exhaust fumes from rush-hour traffic. Wright (1978) reports some deterioration of corrective steering movements on a driving simulator at average blood COHb levels of $6\cdot3 \pm 2\cdot1\%$, but observed no effects on brake reaction time or other driving related skills.

Other studies have found no effect of similar levels of COHb on a range of psychomotor functions (Stewart *et al.* 1970, O'Donnell *et al.* 1971). However, overall the evidence does suggest that exposure to CO is a potential hazard to the driver, with its effects being most marked on the visual processes, and on some central nervous system functions (vigilance and reaction time). The evidence, however, is not unanimous.

2.8. Drivers' assessment of their occupational conditions

Few investigators have addressed the topic of how drivers themselves assess the conditions under which they work; thus we can compose only a rather sketchy picture of those factors which drivers perceive as contributing to fatigue and drowsiness. The following investigators have provided some systematic information: Harris *et al.* (1972) questioned a number of American interstate truck and bus drivers about their attitudes to legislation controlling hours of driving, and about factors relating to fatigue; Edmondson and Oldman (1974), through a relatively unstructured interview, explored the attitudes of a group of 27 English drivers to a variety of potentially stressful aspects of their work; and Fuller (1978) and McDonald (1978) conducted a small survey of the working conditions of drivers in a variety of different types of haulage firm in Ireland. The working hours of both the American and the Irish drivers have been discussed in Section 2.3 above; on average they considerably exceed those of the English sample, who were rarely required to drive more than 5 hours per day. Nearly all the English drivers worked a shift system, alternating between days and nights on a fortnightly basis. A number of the Irish drivers also worked shifts.

Although these investigations are rather heterogeneous in terms of method, scope and sample, they do demonstrate that fatigue and drowsiness are clearly recognized problems of truck drivers. Thus none of the English drivers were prepared to say that fatigue was never a problem, and one-half of them found it at least an occasional problem. Drowsiness was found to be a frequent problem among a clear majority of shiftworkers and an occasional problem among almost all of them. It is more common among shift- than day-workers (being mainly due to inadequate sleep while working on the night-shift); however, it is also a frequent complaint among day-workers.

Fatigue and drowsiness were not well differentiated in these studies. Edmondson and Oldman asked their drivers about their experiences of both fatigue and drowsiness; Harris and his associates asked about fatigue only, and Fuller and McDonald about drowsiness only. It is not clear to what extent these two concepts comprise separate and distinct experiential categories for the drivers; however, there was quite a high level of agreement between the studies concerning about four or five factors which were seen as contributing to a rather undifferentiated combination of drowsiness and fatigue.

The first factor concerns both climate and food and drink; the chief complaint here was having to sit in a hot, poorly ventilated cab, though many drivers also found driving in warm sunny weather or close 'heavy' weather to have a soporific effect. Eating large meals and drinking alcohol were also perceived as causing drowsiness, though the latter was rarely mentioned.

Other aspects of the weather affect the driver by making his task more difficult and demanding; rain, snow, fog and ice are the most obvious and frequently cited examples. Apart from adverse weather conditions, many drivers also find their job taxing and tiring when it involves driving through congested traffic, or when slow driving is enforced by carrying a heavy load.

A third factor contrasts with these conditions which obviously increase the demands placed on the driver's skill: here an absence of traffic and long hours spent on monotonous, uneventful or overfamiliar roads are also conducive to fatigue and drowsiness. Although this kind of situation requires little in the way of action or decision from the driver, what may well be stressful is the combination of boredom and monotony with the need to maintain concentration; and this requirement is perhaps emphasized by the greater speed of travel. It seems that a certain amount of variety in the road and traffic environment is desirable.

A fourth group of problems relate to hours of work and of driving. We have mentioned that drowsiness is an almost universal problem among shiftworkers; many drivers who find it difficult to adapt to shiftwork also complain of feelings of exhaustion and irritability. A large majority of the English sample preferred the day-shift to the night-shift. However, this factor interacts with those we have just mentioned in that most of these drivers preferred driving (rather than working) at night, this being less stressful due to the lack of traffic; nevertheless the greater concentration required at night, and the lack of activity outside the cab were also recognized as making for greater susceptibility to fatigue and drowsiness.

Although relatively few drivers considered long hours of driving *per se* to be a cause of fatigue, some of the English drivers did admit that their driving did deteriorate during a trip; some, on the other hand claimed that their driving improved as they progressively relaxed and increased in confidence. What may be important here is an ability to pace oneself throughout the shift; Edmondson and Oldman (1974) report the interesting idea suggested by some drivers that "whether or not one's driving deteriorated depended to a certain extent on one's ability to monitor one's subjective state and driving efficiency continually, and to alter one's effort and concentration accordingly" (p. 19).

Data produced by the Irish survey make it possible to estimate the extent to which hours of work and hours of driving contribute to a driver's evaluation of his fitness to continue driving. Each driver in this survey was asked, after each of 20 working days, whether he was prepared to drive on, would have preferred to have stopped earlier, or was satisfied with his driving time. Both hours of work and hours of driving had highly significant correlations with the drivers' preferences. Hours of work accounted for slightly more of the variation in drivers' preferences than hours of driving. It is interesting that the contribution of either working or driving time in accounting for the drivers' preferences is fairly low, accounting for 21·5 and 15·6% of the variance respectively (these two factors, it should be noted, are not independent). This suggests that other factors, such as road, weather or traffic conditions or perhaps aspects of the driver's personal and social life, may be more important. The fact that working time made a larger contribution than driving time to the drivers' preferences suggests that other duties are no less arduous than driving, or that the fact of being on duty *per se* makes an important contribution to the driver's evaluation of his fitness to drive. The results is not particularly surprising but it has important implications in, for example, the framing of legislation. Thus the introduction of EEC driving hours regulations into Great Britain involves a shortening of the permitted driving day but no curtailment of the permitted working day (see Table 2.1). This pattern of changes may not be the most appropriate for the improvement of safety. It is worth noting that although on nearly half the days sampled in this survey the drivers expressed satisfaction with the length of time they had driven, on nearly a third of days they would have preferred to have stopped earlier, often by a considerable period of time.

This would seem to be cause for some disquiet concerning the safety of driving conditions typical of that sample.

It is interesting that a quarter of the drivers in the Irish sample expressed no knowledge of the current Irish regulations concerning drivers' hours, and 40% said they were ignorant of the proposed EEC regulations. Regulations thus impinge little on either the consciousness or the working practices of many of these drivers. Of those that did express some opinion on the regulations the vast majority expressed some favourable comment on legislation *per se* (as being fair or needing to be enforced, for example); many were satisfied with the current Irish legislation, though slightly more favoured the EEC legislation as improving conditions or safety. The vast majority of the English and American drivers in their respective surveys were satisfied with the hours that governed their operations; though in the latter study owner-drivers were marginally less in favour of restrictions on drivers' hours.

Although few drivers explicitly stated that high levels of noise and vibration contributed to fatigue, having a comfortable cab is generally thought to be very important. Thus, for example, many of the English drivers said that high levels of cab noise did affect their driving sometimes; though only a few said the same of vibration. Half the Irish sample were dissatisfied with the noise and vibration levels in their cabs. Other inadequacies, concerning the seat, the temperature and ventilation systems and the cab layout, caused many of these drivers discomfort and annoyance, and a few suffered from engine fumes in the cab. A survey by Hewland, Ruder and Finn International Ltd (1976) strongly emphasizes the importance that drivers attach to having adequate seating in their cab (in terms of attenuating vibration, providing postural support, and absorbing perspiration). Although this survey was carried out on behalf of a manufacturer of suspension seats, it is unlikely that their professional interest alone could have engendered the strong relation expressed by the drivers between seating, comfort and health.

Most drivers who suffer from fatigue and drowsiness clearly recognize that their driving skill deteriorates concomitantly. A lowering of the level of concentration and alertness is the most common effect; reactions and decisions also tend to be slower, and there is a greater tendency towards carelessness. One effect of prolonged motorway driving that was reported by some of the English drivers was a loss of the sense of speed; this is probably due to a marked sensory adaptation to the unchanging velocity cues. The most frequent method of counteracting feeling drowsy at the wheel is to stop the vehicle and have a rest or sleep, or to take a walk. Having something to eat or drink is also common.

A number of rather more dramatic and disturbing effects of fatigue and drowsiness were also reported by many drivers in the English survey. These included trucks wandering from one side of the driving lane to another, particularly at night when driving in a fleet of vehicles; driving very close behind another vehicle; wide fluctuations in speed or slowing to a crawl; and failure to see or respond to road-signs. Many drivers reported feelings of being unable to remember passing places they must have passed on a familiar route. Hallucinations had been experienced by 29% of the drivers, often involving 'seeing' objects on the road, and even taking avoiding action. Many hallucinations were associated with fog or lack of visual stimulation at night. Very similar experiences were reported by a group of long-distance and local drivers in the southern USA (McFarland and Mosely 1954); many of these drivers had experienced an inability to interpret the meaning of signs—for example a red stop light or a posted speed limit may mean nothing to the driver. Also, loss of a sense of time and place can occur with failure to recognize familiar places or 'recognizing' previously unvisited places. All of the 33 long-distance drivers had suffered from 'hypnagogic hallucinations' which the authors interpret as terms of the wish-fulfilment of a strong desire to stop, which is prevented in reality by the necessity of keeping to schedule. A particularly dramatic case is as follows:

I used to drive on routes in the Southwest and on one route the road was very flat and straight for long distances. I used to drive at night and my eyes would get tired. I would close my eyes and count 1, 2, 3, 4, 5, then open them and check my position and then close them and count again. One time I had counted up to four and saw suddenly sitting on the road a very large colonial mansion with an elaborate colonnade in front. I didn't know how it got on the road, but I turned quickly to the left to avoid driving into the building and woke up with my truck turning over in a gully about 20 feet down below the level of the road. It was a miracle that I was alive. Then I realised that I had gone to sleep at the wheel! (McFarland and Mosely 1954; p. 126).

2.9. Aspects of the career of the truck driver

The typical career structure of the truck driver in Britain, according to Hollowell (1968), is shown, in very general terms, in Figure 2.8, though the considerable variability in career patterns which he found should be emphasized. 'Shunting' refers

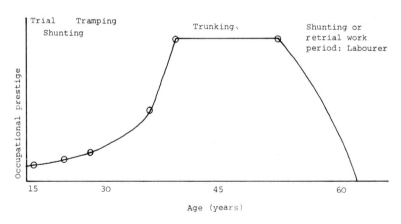

Figure 2.8.   A model of the truck driver's career. After Hollowell (1968).

to almost any kind of local delivery driving and usually involves loading and unloading the vehicle. A 'trunker' normally drives a fixed regular route, often on a night-shift or on an alternating shift system; he usually returns home every day. In 'tramping' there is no set route; typically the driver will not know in advance where his week's work will take him, but will take a load from one depot to another on the first day out, pick up a load there and take it to a third point and so on. Thus the tramper may only be home at weekends, if that. His work often involves loading and unloading, and delivering goods to customers, which can be physically arduous, and he frequently has to find his loads himself. The typical career pattern explored by Hollowell is roughly as follows: most truck drivers come to that occupation either from a non-driving mobile occupation or an occupation from which driving is a 'visible' occupation. Starting in a local delivery job the tendency is, once experience has been gained, to move into long-distance work, which is more 'adventurous' and less monotonous than local work and when an increase in earnings is probably necessary due to family responsibilities. As the driver gets older and can no longer so easily handle the heavy work involved in tramping, and due to the considerable strains of maintaining a family or home community life, the greater regularity of trunking is likely to be preferred. However, it is likely that the social and psychological strains of shiftwork may induce him to give that up, and generally speaking the truck driver will either go back to being a shunter, without any heavy loading, which involves a loss in occupational prestige and earnings; or he might take up another job altogether, which may involve a range of options, from labouring to being lucky enough to get a junior managerial job.

A rough breakdown of the major occupational divisions in the haulage industry is given by Carré and Hamelin (1978). There is a major distinction between those who own their own vehicle, or several vehicles, and those who work for wages. While the latter have traditionally tended to aspire to owning their own vehicle or small haulage business, this has become decreasingly the case. The owner-driver's situation appears less enviable as the growth of subcontracting tends to make his working conditions more like those of the waged driver, and as the increasing capital costs of the vehicle makes for greater indebtedness, longer paying-off times and greater financial insecurity. The status divisions within the industry favour long-distance and international haulage over regional or middle-distance haulage, which is in turn valued higher than local haulage or public works. However, within this hierarchy there is a well developed consciousness of what makes for a good or bad firm or job. This involves a variety of factors, including wages and conditions of work, the economic stability of the firm and the attitude of its management to its drivers, as well as more technical criteria such as the type of freight (which has implications for loading work), the type of vehicle (where, for example, power and speed are important), and the organization of work shifts. There is a further social division—between young and old drivers—and this complements the career evolution outlined by Hollowell above. Older drivers accuse younger 'cowboys' of undermining the norms and conditions of work enjoyed by the rest by working longer and faster, seduced by the extra bonuses they can earn. Younger drivers, it is felt, are less dependent on returning home every week, resent less the privations of work, and are less aware of the consequences of their work for safety; whereas the older drivers are not prepared to push productivity so that it costs more in fatigue, stress and risk than it brings back in money.

Almost universally, truck drivers seem to value their job for the autonomy that it gives them. This has been remarked on by various researchers (Hollowell 1968, Hamelin 1975, Lille 1976). Truck driving is, in Hollowell's terms, an 'open socio-technical system' which implies considerable freedom from organizational constraint, and also a long 'job cycle time' (thus a trunker's cycle time will normally be a whole shift, involving a completed return journey; a tramper's cycle time may be a whole week; on the other hand an assembly line worker may have a cycle time lasting only a few minutes before he repeats the same sequence of operations). It is this feature of the job which, Hollowell argues, explains the high level of job satisfaction among lorry drivers and the persistence in remaining in the occupation. This feeling of autonomy also goes some way towards explaining why many drivers feel strongly opposed to attempts to regulate their working conditions, as witnessed, for example, by the major campaigns against the compulsory introduction of tachographs in Britain. This strong sense of freedom from managerial control persists even in jobs which are highly regular and predictable, and where the timing of the driver's progress along the route is precise and unvarying. This should not, however, be seen as a contradiction.

There is some variety in attitudes towards regulations, as has been pointed out earlier. While such attitudes have tended to be corporatist in character (reflecting the perceived interests of the industry as a whole) and not favouring restrictive regulations, it is also true that drivers do frequently favour a strict control of working hours, and they certainly resent their personal liability for contravening the regulations when their working rhythms are seen as being imposed upon them from outside. Many would like to see their working hours become increasingly aligned with the norms of other workers (Carré and Hamelin 1978). It is an industry where the collective enforcement of improved working norms against managerial interests has not often proven very successful (as the evidence on working hours demonstrates). Whether this might make for an increasingly favourable attitude towards regulations is not clear; certainly, in many countries regulations have been less than successful in influencing norms of work.

A good description of the image that many truck drivers have of their occupation is provided by Lille's (1976) report on French *grands routiers*, part of the summary of which translates as follows:

This image is of an occupation in the practice of which there is a certain freedom, a certain realization of the personality, but which, in return, leaves little freedom or leisure outside of it. It is of an occupation that is hard, exacting, and which values those who are capable of performing it. It is of an occupation which one chooses, a useful occupation, a vocation. But it is also of an occupation that is paid in wages, where the exploitation of individuals is pushed to limits which are often dangerous, without the remuneration being adjusted for these excessive efforts, without the deterioration in health being compensated by the recognition of certain professional illnesses, nor being prematurely worn out being compensated by an early retirement. Moreover, the wage-earners seem to attribute far less of the responsibility for this situation to their boss than to the whole system, and notably to the government, to the administration, to the SNCF.

2.10. *Summary and conclusions*

While regulations and collective agreements controlling drivers' hours of driving, work, and rest are common throughout the world, the form that these take does vary quite significantly. And while the elimination of fatigue is frequently invoked in the rationale for such regulations, it is clear that the economic and commercial consequences of such regulations are also important influences. The attitudes of drivers and hauliers to such regulations will also be conditioned by a variety of considerations, including the desire to have conditions of work and life that approximate to those enjoyed by other workers, and concern for the commercial viability and profitability of the industry in which they work. These considerations reinforce the need for an independent evaluation of such regulations in relation to safety.

It is clear from surveys in various European countries and in the USA that very long hours of work and of driving are the norm in the haulage industry; and that regulations on drivers' hours are not enforced, or are ineffectively enforced, in many countries. Rather, the limitations on drivers' hours appear to reflect more a balance solely between work time and time for the reconstitution of the driver's labour power through rest, food, sleep and hygiene requirements; though time and facilities for the latter are frequently and regularly severely curtailed and qualitatively inadequate. Leisure and domestic life thus do not enter into the time equation during the working week. Long-distance and international drivers appear to work and drive the longest hours. Geographical separation from home and base means that other conditions of work and rest also tend to be inadequate, including facilities en route, at the customer's depot, and the environment of the truck cab for sleep.

Shiftworking is also common in the haulage industry. A wide range of shift systems are worked; some are highly irregular and unpredictable (some firms want to keep loading and dispatching continuously, so the driver's shift can start at any time during the 24-hour cycle); others are more predictable but arduous ('sleeper' operations involve two drivers alternating driving and sleeping for the working week); and various forms of alternation, rotation or permanence of shift exist. Shiftworking involves taking at least some sleep during the day, which therefore tends to be shorter, more interrupted and of worse quality. Working at night thus involves the stresses of working during the low phase of circadian arousal, and deprivation of sleep, as well as the effects of prolonged work. Under such circumstances one might expect a decrement in performance, particularly in vigilance, and difficulty in keeping awake. Shiftwork has been associated with a range of adverse effects on health and well-being. Many workers do, however, appreciate the advantages of shiftwork, particularly extra money and free time during the day. On the negative side there is the likely disruption to domestic and social life, and disturbances of sleep, digestion and mood, which may on occasions be serious. No reliable way of predicting individual differences in adjustment to shiftwork has been found. The combination of a number of different aspects of the truck driver's working conditions may be implicated in higher than expected incidence of several digestive and circulatory disorders (haemorrhoids, peptic ulcers, appendicitis and nervous stomach) which have been found in truck drivers when compared to other occupations. Such factors include working shifts, an irregular and unbalanced diet, the need to maintain a constant seated posture, whole body vibration, lack of exercise and the stress of the job. The precise role of these factors in the aetiology of these disorders can only be speculated upon. Digestive problems may also be exacerbated by a high level of consumption of coffee and cigarettes, and the latter may underlie a high mortality rate for lung cancer among goods vehicle drivers.

Noise levels at the driver's ear in truck cabs tend to be underestimated by standard methods of measurement of in-cab noise. Noise levels in the cab vary between different cabs and operating conditions, and also depend on whether the windows are open and whether a CB or other radio is being used. Noise levels exceeding 90 dB(A) are not unusual, and those exceeding 85 dB(A) are common in standard truck cabs under normal conditions. The effectiveness of techniques for the sound insulation of the truck cab and for quietening the engine have been well demonstrated, and the potential exists for the considerable improvement of truck noise levels. The legislative trend appears to be towards insisting on quieter vehicles. Such improvements are necessary to avert the possibility of damage to the hearing of truck drivers as personal forms of hearing protection are not suitable for driving. In-cab noise levels may tend to mask auditory signals from other road users, and may be detrimental to the driver's mood and performance. These effects may be manifest in feelings of annoyance, discomfort and fatigue, and in the quality of interactions with other road users; based on extrapolations from laboratory studies, complex decisions in traffic and vigilance (particularly in the periphery of attention and under time pressure) may be the most vulnerable aspects of driving performance. There is little reliable evidence that levels of infrasound typical of those found in trucks have any specific damaging effects.

The frequencies of maximal human sensitivity to whole-body vibration tend to coincide approximately with the vibration resonance frequencies of most trucks, particularly in the vertical axis, which together with pitch is the most severe vibration mode impinging on the truck driver. Again, levels of vibration at the driver's seat vary considerably in different trucks, on different roads and at different speeds; however, on typical North American roads vibration levels have been found to vary between below the 4-hour Fatigue Decreased Proficiency boundary of the ISO and below the 1-hour boundary. While the ISO limits have been criticized on a number of grounds, they have also been defended as the best available summary of the evidence on adverse effects of whole-body vibration. At levels experienced by many drivers, some direct mechanical effects of vibration on visual acuity (where vision through cab mirrors is likely to be most affected) and manual control (where there may be some loss of fine steering control) may be expected. There is little evidence concerning the effects of prolonged exposure to vibration on performance, but the most important effect is likely to be not a direct mechanical effect, but the result of prolonged exposure to discomfort, physiological strain on fatigue and motivation to perform well. Such processes are not easily simulated in the laboratory. The other direct effect of whole-body vibration on the driver's well-being is the exacerbation of back and lumbar spinal disorders, where a combination of postural stress, vibration, muscular effort, and shock or impact are likely to have contributed to the high incidence of back problems which has been found among truck drivers. Measures to reduce the levels of vibration affecting the driver have been technically feasible for many years; they involve either the reduction of the natural resonances of the truck suspension systems at the chassis, or in isolating the truck cab or driver's seat. The provision of seating with good lumbar support is necessary to protect the driver's back.

Depending on the climate and time of year, extremes of temperature may be a threat to the driver's well-being and safe performance, though with efficient heaters being standard equipment and more common than full air-conditioning systems, the greatest problem is probably high temperatures. Much will depend on the state of acclimatization of the driver, but it is likely that his performance will begin to suffer within a few degrees of his preferred temperature, when he may also tend to become drowsy or irritable. High radiant temperatures from the sun are likely to be a major source of discomfort. To prevent discomfort from perspiration, the driver's seat should be covered with absorbant material.

The evidence that carbon monoxide is a possible danger to the driver is mixed. There is no strong evidence that driving performance is likely to be worse with blood carboxyhaemoglobin levels typical of those produced either by smoking or by heavy traffic. However, in any situation which might produce an unusually high level of carboxyhaemoglobin, there may be an impairment in performance, particularly in visual functions.

Drivers themselves recognize that a wide range of factors contribute to feelings of fatigue and drowsiness and deteriorating driving performance. Thus hot weather or high cab temperature and the period after a large meal make for drowsiness; bad weather, congested traffic or driving a slow vehicle appear to make the driver's job more taxing; long hours on monotonous roads cause problems in maintaining alertness; and night driving is strongly associated with drowsiness and difficulty in maintaining concentration. Hours of work are more important than hours of driving on their own in predicting fatigue. The most effective way to counteract fatigue and drowsiness is to stop and have a break or a short sleep. Thus, it is important that drivers should have the flexibility to do this when appropriate.

Truck drivers value their job for the sense of autonomy which it brings them, and this may be some compensation for the extreme hours of work which are the norm in their industry. However, the stresses of work do seem to structure the career of the driver, and the extreme privations that many younger drivers are prepared to undergo are seen by their elder colleagues to be undermining conditions of work in the industry as a whole.

Thus it is abundantly clear that there is a problem of fatigue among truck drivers. This problem involves not only long hours of driving but the whole pattern of work, rest and leisure throughout the 24-hour and weekly cycles. It involves facilities for rest and sleep and for food and personal care; it involves the physical environment in which the truck driver works, and it involves the road and traffic environment.

It is now appropriate to turn to try to estimate the magnitude of the problem in terms of safety, and then to look at the processes of fatigue in terms of the breakdown of skill, physiological degradation, and the subjective experience of the driver.

Chapter 3
Accidents

3.1. Introduction

One of the main criteria of safety is, almost by definition, the occurrence of accidents. This chapter assesses the evidence that 'fatiguing conditions' are associated with higher rates of driving accidents than might otherwise occur; and, reciprocally, that a significant proportion of accidents can be attributed to 'fatigue', falling asleep at the wheel and related factors. First of all, general levels of safety in truck driving will be discussed, both from the point of view of the truck driver himself in comparison to other workers, as well as in terms of the contribution of trucks to the severity of accidents in which they are involved, and the relative accident rates of other types of vehicles. The third section of this chapter is concerned with the causation of accidents and attempts to assess the contribution of 'fatigue' and falling asleep to the accidents of truck drivers, and will also discuss the nature of those accidents. Finally, the relationship between time and accidents will be discussed, both in terms of the number of hours spent driving before an accident, and the time of day at which accidents occur.

3.2. General levels of safety in the haulage industry

How dangerous is truck driving as a profession? How does the risk of death or injury compare with other jobs? In the USA it ranks among the top 10 industries in the private sector for its rate of lost workdays due to industrial injury and illness (this is based on a three-digit Standard Industrial Classification in which there are several hundred categories). In this respect it is comparable to shipbuilding, iron and steel foundries and sawmills, and comes slightly below, for example, water transportation services, logging, and roofing and sheetmetal work. Its level of lost workdays is approximately three times the average of the private sector as a whole. Although well above average, trucking is not quite as high in the hierarchy of dangerous industries in rates of injury as it is in relation to the number of days lost; this would seem to imply that injuries tend to be more severe in trucking than in other industries (see US Department of Labor 1976). These figures do not, however, tell us about rates of death.

Occupational mortality statistics indicate a high mortality ratio (mortality compared to population norms) among goods vehicle drivers. The greatest part of this excess mortality is accounted for by deaths in motor vehicle traffic accidents, the rate of which is virtually twice that would be expected on the basis of population norms. Deaths in non–traffic accidents, although smaller in number, also show a very high mortality ratio for goods vehicle drivers. This excessive rate of accidental death is most evident in the 25–34 years age group. The only other statistically significant factor contributing to the higher occupational mortality of goods vehicle drivers (after correction for social class) is cancer of the lung, which is attributed to a possibly higher rate of smoking in this group (Office of Population Censuses and Surveys 1978).

There appears to have been a steady increase in the mortality ratio of goods vehicle drivers over the half century following 1920, in Great Britain at least. Whereas in the early 1920s only syphilis and cancer showed an excess mortality for motor vehicle drivers (all other causes of death including accidents were below population norms), 10 years later accidents had exceeded the norm and the increase was much greater in the younger age group. Since then the mortality ratio of goods vehicle drivers has steadily increased until the last available figures for the early 1970s (see Table 3.1). Since then there appears to have been a gradual decline in the rate of death of drivers of heavy goods vehicles as measured by road accident statistics. Figure 3.1 shows the rate of death per 100 million kilometres for HGV drivers during this period. Although the mileage of these vehicles has increased during this period, the fluctuations in the

Table 3.1.   *Standard mortality ratios for goods vehicle drivers between 1921–23 and 1970–72.*

	All causes of death	Death in accidents	Deaths in motor vehicle accidents
1921–23[a]	84	86[b]	
1930–32	76[b]	103	
1949–53	91		191
1959–63	101		154
1970–72	111		194

[a] Includes both passenger and goods vehicle drivers.
[b] Approximate.
Source: McDonald (1981)

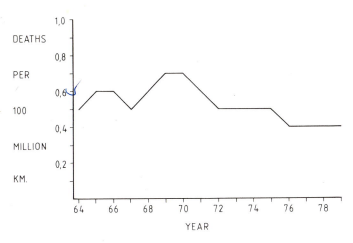

Figure 3.1. *Road accident death rates of HGV drivers, 1964–79.* Source: *McDonald (1981).*

accident numbers are proportionally much greater than the mileage fluctuations. However, this does not take into account any changes there may have been in the numbers of heavy goods vehicle drivers, so it is impossible to predict what has happened to the comparative occupational mortality figures during this period (McDonald 1981).

If we turn now to consider the involvement of goods vehicles in the total pattern of road accidents, the evidence is clear that goods vehicles are involved in a large proportion of accidents, particularly fatal accidents, and that the heavier the vehicle the more accentuated is this pattern. Thus, for example, in Great Britain in 1981 heavy goods vehicles (over $1\frac{1}{2}$ tons unladen weight) accounted for 9% of all vehicle involvements in fatal accidents, 4% of involvements in serious accidents and 3% in slight accidents (Department of Transport 1982); and in 1973 it is estimated that goods vehicles over $4\frac{1}{2}$ tons unladen, while comprising less than half of the haulage fleet, account for 59% of all HGV accidents and 70% of fatal accidents. Equivalent figures for Ireland in 1979 show an involvement rate of goods vehicles in fatal accidents of 15% and in personal injury accidents 10% of all vehicle involvements. Goods vehicles over 5 tons unladen weight account for 32% of accidents while making up only 15% of the haulage fleet (An Foras Forbartha 1976, 1980). There are two factors which together largely account for this pattern and these relate to the relative mass of different types of vehicle and their annual average mileage.

The loaded weight of a large articulated truck may be some 30 or 40 times the weight of a car; thus it is not surprising that impacts between trucks and cars result in very high rates of fatality and serious injury to the occupants of the latter. Grime and Hutchinson (1979) have shown that it is the mass ratio between vehicles rather than the mass *per se* which is the major determinant of driver injury in head-on collisions. This is well illustrated in Table 3·2, which was compiled from reports of

Table 3.2.   *Fatal crash involvement rate per 100 000 years registered by type of vehicle,*
Maryland, 1970–71.

Type of involvement	Motorcycles	Cars	Non-trailer trucks	Tractor-trailers
Single vehicle— occupant death[a]	35·1	7·4	8·4	25·2
Multiple vehicle— occupant death[a]	49·9	7·8	7·3	4·2
Total in which occupant died	85·0	15·2	15·7	29·4
Multiple vehicle— no occupant death[b]	1·8	5·6	14·6	87·9
Pedestrian death (by size of striking vehicle)	3·7	6·6	8·0	29·3
Total fatal crash involvement per 100 000 years registered	90·5	27·4	38·3	146·6
No. of fatal crash- involved vehicles	49	885	173	35
No. of years registered[c]	54 120	3 227 214	451 126	23 882

[a] Death occurred to occupant (including driver, or any rider in the case of a motorcycle) of the vehicle described at the head of the columns.

[b] No death to occupants of this vehicle but death occurred in another vehicle involved in a crash with this vehicle.

[c] No. of years registered reflects the cumulated total of registered vehicles of this type, and hence the number exposed to risk.

After Robertson and Baker (1975).

1440 fatal crashes in Maryland, USA (Robertson and Baker 1975); while tractor-trailer units have very low rates of crash involvements in which a tractor-trailer occupant was killed, they have very high rates of involvements in which occupants of other vehicles were killed. In crashes between trucks and large cars, death occurred three times as often in the cars as the trucks, and between small cars and trucks the ratio was six to one. A study by Baker *et al.* (1975) of 150 crashes involving tractor-trailer units found, in similar vein, that when cars were involved in these crashes, the vehicle occupant death rate was almost ten times as high in the car as in the tractor-trailer (56% as compared to 6%); indeed tractor-trailer occupants were only killed in such crashes if in addition to striking the car the tractor-trailer either overturned, went down an embankment or struck something like an embankment, abutment or bus. Gissane and Bull (1973) found even more extreme figures in their sample of car/lorry collisions: 224 deaths of car occupants were matched by five deaths of lorry occupants. Pedestrians are also greatly at risk from large trucks, as is demonstrated by the Robertson and Baker study (see Table 3.2). This figure also demonstrates the relatively high rates of death to tractor-trailer occupants in single-vehicle accidents; and this finding is confirmed by Jones (1976).

As well as simply the mass of the truck involved, other structural features of the vehicle contribute to the severity of the accident. Both Robertson and Baker (1975) and Baker *et al.* (1975) emphasize the need for better energy management in crashes (in the former study 88·5% of vehicle occupants survived crashes in which another occupant of the same vehicle died).

Current legislation on maximum length and weight limits encourages making tractor units out of lighter materials and with cab-over-engine designs which lack protective frontal projections. Baker *et al.* (1975) suggest that units for length and weight should be exclusive of the tractor, to counteract the present tendency to sacrifice the safety of occupants in the interest of greater payloads. Standards to minimize roof crush, side intrusion and damage to occupants from impact with interior structures have at present been set only for passenger cars. Given the provision of a stronger cab structure, the driver then requires protection against serious injuries resulting from striking the steering wheel or other hard surfaces in the cab; such protection can only be provided by appropriately designed seat-belt systems, and their effectiveness has been amply demonstrated (Johannessen 1970, Grattan and Hobbs 1978, Rüter and Hortschick 1979, and Högström and Svenson 1980). Many occupants of tractor units are trapped for long periods, indicating the need for better escape and extrication provisions. The severe consequences of underriding the rear or side of a trailer by a car, particularly a small car, could be ameliorated by the provision of improved rear and side protectors. Neilson *et al.* (1979) suggest some additional measures that might reduce the incidence of HGV accidents. These include improvement in tyre wet grip, braking system performance, loading and load-securing techniques, and design features affecting roll-over.

While the weight of heavy trucks can largely account for the high proportions of fatality or other injury, particularly to other road users, in accidents in which these vehicles are involved, their high overall accident involvement rate can be accounted for by the much higher mileage of the heavier trucks. Thus, for example, Gissane and Bull (1973) suggest that the average annual mileage of a truck in the unladen weight range 1½–3 tons is 12 000 miles, while for a truck over 10 tons the average is 40 000 miles. Sexton (1967) reports a similar pattern in his survey of the Irish haulage industry. Table 3.3 shows the involvement rate per hundred million kilometres

Table 3.3. *Vehicle involvement rates per hundred million vehicle kilometres by class of vehicle and severity, Great Britain, 1981.*

	Severity of accidents		
	Fatal	Fatal or serious	All severities
Goods vehicles over 1½ tons unladen weight	4·0	25	70
Goods vehicles under 1½ tons unladen weight	2·4	27	100
Cars	2·4	31	119
All motor vehicles	3·1	41	148

Source: Department of Transport 1982.

travelled of various classes of vehicle in Great Britain in 1981. Thus, while heavy
goods vehicles have a lower overall involvement rate than the average for all motor
vehicles, their involvement rate in fatal accidents is higher than average. This pattern
of total accident involvement rates is also very similar in Ireland (An Foras Forbartha
1976; p. 4) and Sweden (Mohlin and Kritz 1972).

When the number of casualties rather than the number of fatal accidents is taken as
the basis for statistical comparisons, the damage inflicted by HGVs is even more
starkly displayed. A report in the *Sunday Times* of 27 June 1976 showed that the
heaviest goods vehicles (over 10 tons unladen weight) account for over three times the
fatalities per vehicle per mile of light goods vehicles and cars. When adjustments are
made for the contribution of the weight of each vehicle to the severity of accident, the
fatality rate attributable to the involvement of HGVs in accidents is further inflated
(Wardroper 1976). A small study by Ruffel-Smith (1970) obtained a similar pattern
by investigating the other vehicle involved in collisions in which a car occupant was
killed. When adjusted for their respective mileages HGVs had ten times the
involvement of cars and light vans in the approximately 300 fatalities which he
investigated.

Thus, to summarize this section, even on the most conservative estimate, truck
driving has an occupational mortality rate comparable to the highest in manufacturing and construction industry. Truck drivers themselves are particularly at risk in
single-vehicle accidents; in multi-vehicle accidents involving trucks the occupants of
other vehicles are especially vulnerable to fatal injury, and the same is true for
pedestrians involved in accidents with trucks. Although HGVs have a lower
involvement rate of cars and light vans in the approximately 300 fatalities which he
their involvement rate in fatal accidents is higher than the mean for all vehicles, and
they account for a disproportionate number of deaths even taking into account their
great mileage.

3.3. Fatigue, falling asleep and the nature of accidents

A rough indication of the incidence of falling asleep at the wheel among the general
driving population is provided by two studies, one by Prokop and Prokop (1955), the
other by Tilley *et al.* (1973). The former study found that of 569 drivers in motorway
cafes who filled out a questionnaire, 18% admitted that they had fallen asleep at the
wheel at some time. The latter study gave its questionnaire to driving licence renewal
applicants. Sixty four per cent of the 1500 respondents reported having become
drowsy while driving at least at some time; of these, nearly 10% said that drowsiness
or falling asleep has caused one or more near-accidents; and 10% of this group said
they had had at least one accident under these circumstances. Truck and bus drivers
themselves attribute a larger proportion of the accidents in which they are involved to
factors related to fatigue. Fifty per cent of the drivers surveyed by Mackie and Miller

(1978) had had at least one accident; of these, 3% were attributed to sleep or drowsiness, 2% to boredom, and 1% each to too much loading work and driving too long.

Positive identification in accident statistics of fatigue or falling asleep as contributory factors is likely to underestimate their full contribution: the driver himself is usually the main source of such information and he will tend to underplay his contribution to the accident and overestimate his vehicle's (Baker 1967). There are a few studies, however, which do suggest that 'fatigue' and falling asleep at the wheel are important contributors to truck drivers' accidents, though the extent to which these factors have been implicated varies greatly. In Sweden, for example, Mohlin and Kritz (1972) studied accidents involving heavy articulated vehicles in which a defect in that vehicle or an impairment in the drivers' capabilties could be shown to have contributed to the accident; 20% of such accidents involved driver fatigue, and this was just over 1% of the total accidents involving heavy articulated trucks. In the USA, the Bureau of Motor Carrier Safety (1971) followed up accident reports by haulage firms which suggested that the driver's condition may have contributed to the accident; of the 400 drivers concerned, 76% had fallen asleep at the wheel. In two samples of accident reports selected independently of the driver's condition, Harris (1977) found that in about 7% of single-vehicle accidents and 5% of other vehicle accidents the driver had been registered as 'dozed at the wheel'. Finally, Brown (1967 a) quotes UK accident statistics for 1965 which suggest that while 'fatigue' is not a major cause of accidents (in comparison with other identifiable causes) it is a more prevalent factor amongst goods vehicle drivers than other drivers (see Table 3.4). More recent accident data in the UK are not classified according to such contributory factors.

Quite apart from these cases where 'fatigue' or falling asleep have been identified as the major contributing factor to the accident, there are a much larger number of cases where the driver has not responded adequately to the situation; and a range of factors in the 'fatigue' complex may well have contributed to such an impairment in the driver's functioning. Chapter 4 will consider the evidence that conditions such as

Table 3.4. Fatal and serious road accidents in the UK during 1965 (number per million vehicles at risk) by 'cause' and class of vehicle.

Taxation class of vehicle	Driver fatigued	Driver intoxicated	Learner driver	Excessive speed	Slippery road	Dog in road
Motor bicycles	40	20	5101	1296	4741	155
Cars and taxis	39	55	225	593	3190	20
Public service vehicles	21	31	186	660	16691	72
Goods vehicles	51	85	213	68	6258	30
All vehicles	**40**	**54**	**866**	**697**	**3904**	**40**

After Brown (1967a).

prolonged driving, inadequate sleep, or disturbances in sleep–waking rhythms may lead to a deterioration in the driver's skill. Accident statistics can give little useful information in this respect; however, the high proportion of accidents attributable to inattention or failure to look may provide one reference point for such considerations. For example, Starks (1957) found that 14% of nearly 10 000 accidents on the Pennsylvania turnpike were attributable to driver inattention, and Clayton (1972) found that 'failure to look' accounted for 28·5% of identifiable accidentogenic driver errors in his sample of nearly 300 accidents. A significant involvement of fatigue in driver inattention accidents was found by Shinar *et al.* (1980), in a post-accident investigation and analysis. The role of fatigue was also particularly prevalent in, not surprisingly, 'critical non-performance' accidents such as falling asleep at the wheel. Fatigue was implicated in each of these categories four or five times out of the 420 accidents analysed and in 18 of the accidents altogether (which is slightly over 4%).

Thus in terms of various ways of attributing driver states as causes of accidents, fatigue, drowsiness and related states have been associated with anything between roughly 1% and 7% of accidents. None of these figures are reliable however, and perhaps the most important aspect of fatigue they neglect entirely is the possible contribution of fatigue in a rather subtle but pervasive manner to a wide range of accidents which are more obviously attributable to more immediate and tangible factors. For example, Hills (1980), in his analysis of the visual requirements of driving, emphasizes the great scope for error in visual search, expectancy and perceptual judgement, and the importance of these in accident causation. One can infer from this that any factor which might affect, even slightly, the efficiency with which these functions are performed could have an influence which is indeed subtle and pervasive in accident causation.

Different contributory factors can be identified to a greater or lesser extent with certain types of accident. Perhaps the classic falling asleep at the wheel accident involves no other vehicle. Estimates again vary as to the proportion of these being attributable to falling asleep, Baker (1967) finding that one-quarter of single vehicle accidents involved reports of 'driver asleep', while Harris's figure for truck accidents (mentioned above) was 7%.

There is evidence that commercial vehicles, particularly large trucks, have lower rates of involvement in single vehicle accidents than other types of vehicle (Baker 1967, 1968, Chatfield and Hosea 1972); though this does not necessarily mean that fatigue and falling asleep are correspondingly less important contributory factors. The vast majority of the fatigue and falling asleep accidents in the US and Swedish studies of truck accidents quoted above were single-vehicle accidents.

When more than one vehicle is involved in an accident it is rear-end collisions that produce the greatest number of reports of 'driver asleep' or 'fatigued', though not to the same extent as in single-vehicle accidents (Chatfield and Hosea 1972, Kishida 1981). Numerous studies have shown that trucks have a disproportionately high involvement rate in rear-end collisions in relation to their contribution to total traffic (Starks 1957, Farachi *et al.* 1971, Hosea and Chatfield 1972, Chatfield and Hosea 1972,

Baker *et al.* 1975). Chatfield and Hosea report that drivers of tractor-trailer combinations are responsible for an excessive proportion of these collisions. There are several reasons for this pattern of accident involvement apart from the contribution of fatigue and falling asleep. The relatively low power-to-weight ratio of most heavy trucks means that they are unable to maintain speed on an upgrade and are thus likely to be struck from behind. Trucks also have poor braking performance relative to other vehicles—Haddon (1971) reports that braking performance of trucks is commonly two to three times worse than that of cars. This is why on turnpike roads in the USA trucks are involved as the striking vehicle in rear-end collisions more than twice as often as expected (Baker *et al.* 1975; p. 8). Driving within tight schedules may also contribute to this kind of accident and may explain why McFarland (1957) found that following too closely was one of the most frequent contributing factors to near-accidents in his study of long-haul bus operations.

In summary, therefore, fatigue and falling asleep have been positively identified as contributing to a certain proportion of truck accidents, though no precise estimate of the magnitude of the problem is possible. These factors account for the vast majority of cases where the truck driver's capabilities have been found to be impaired in some way, and this may be between 1 and 7% of total truck-driving accidents. It is likely that fatigue and drowsiness will tend to be under-reported as contributed to accidents because the driver himself is the major source of this kind of information; and certainly only fairly extreme cases will tend to be reported in this way. Fatigue and falling asleep have been particularly implicated in single-vehicle accidents and rear-end collisions.

3.4. *Accidents and the time factor—hours of work and time of day*

One of the most surprising findings to researchers in the fatigue area comes from a study by Potts (reported in McFarland and Mosely 1954). Recording the number of near accidents over 5000 miles of commercial truck runs in the USA, Potts found a dramatic decline in the numbers of near-accidents during each shift of driving (see Figure 3.2). Twenty three per cent of near-accidents occurred in the first hour, 46% in the first two hours and only 2% in the ninth hour of each trip, decisively turning a conventional 'fatigue' account of prolonged driving on its head. Several suggestions can be advanced to account for these findings; however, none of them are very satisfactory or conclusive. The 20 trips over which these data were collected were apparently all around eight or nine hours' duration, so there is no bias towards short trips. However, McFarland and Mosely report no data on traffic density; it is quite feasible to suggest that exposure to risk of near-accident is a function of traffic density, which may well have given a bias towards the earlier parts of the trip. The observer (an experienced truck driver himself) was working with the following definition of near-accident: 'emergency situations or critical incidents which could easily have led

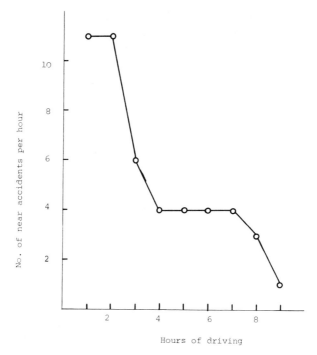

No. of near accidents per hour

Hours of driving

Figure 3.2.   Number of near-accidents incurred per hour of driving.
After McFarland and Moseley (1954).

to an accident' (p. 240). However, it is unlikely that the occurrence of near-accidents is related in any simple way to the occurrence of accidents, and when the focus is on accidents to which some 'fatigue' dimensions may have contributed the relationship may be tenuous indeed. Such factors as the sort of subjective safety margins within which the driver is operating and the accuracy of his appraisal of his own capabilities obviously affects this driving style; the occurrence of near-accidents does not necessarily reflect the level of the driver's capabilities, or any putative criterion of 'objective accident risk'. Road and traffic situations which throw up a high rate of near-accidents are not necessarily those where a 'fatigued' driver is most susceptible to having an accident. The majority of near-accidents in Potts's study involved the driver in question making the assumption that the other vehicle or vehicles would take action to avoid a collision. A more 'defensive' driving strategy was possibly employed as the trips progressed. Among some drivers two factors, coming to work improperly rested, and emotional problems resulting from the drivers' personal or company relations, may have contributed to the obtained pattern of near-accidents.

    In contrast to these results are the findings of Harris *et al.* (1972). Working from the accident records of two large haulage companies and a bus company in the USA, they compared the number of accidents reported per hour of driving before the accident, to the expected distribution of accidents based upon the distribution of different lengths of driving period. Data from the larger trucking company are shown in Figure 3.3 in the form of the ratio of obtained to expected accident frequencies. Between three and six hours after the start of driving this ratio is slightly smaller than unity, thereafter increasing, showing a higher than expected rate of accidents between the seventh and the tenth hours of driving; the differences between the obtained and expected proportions of accidents are statistically significant. It should be noted, however, that the ratios for the ninth and tenth hours of driving are based on comparatively small numbers of accidents and driving trips, so while being the most interesting for interpretation, they are the least reliable of the figures. A similar analysis for the other haulage company showed no significant results; this was also true for the bus company when all the accident and driving data were treated together. However, when daylight driving was compared to night and younger drivers compared to older, significant differences were found between each member of these pairs as a function of time driving, with night driving and older drivers showing an increasing ratio of obtained to expected proportions of accidents as a function of time driving (see Figure 3.4). Older drivers, it should be mentioned in passing, have much lower overall rates of accident, despite being possibly more susceptible to the effects of prolonged driving.

This general pattern relating hours of driving to increased accident risk has been confirmed by a later study reported both by Harris (1977) and Mackie and Miller (1978), which was carried out on behalf of the US Bureau of Motor Carrier Safety. Using a revised accident report form which for the first time obtained information

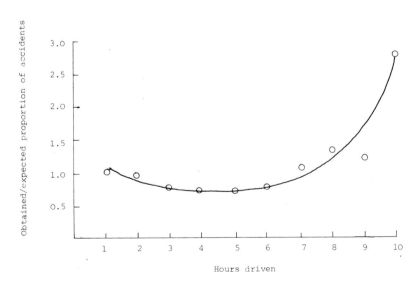

Figure 3.3. *Ratio of obtained to expected proportions of accidents as a function of hours driven.*
After Harris et al. (1972).

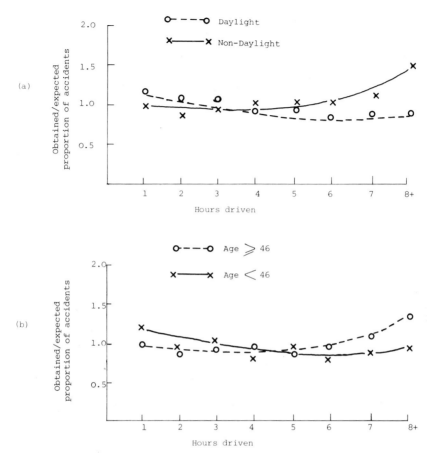

Figure 3.4.   Ratios of obtained to expected proportion of accidents as a function of hours driven; (a) for daylight vs. non-daylight accidents; and (b) for older vs. younger drivers. After Harris et al. (1972).

about hours of driving, this study was based on a much larger sample than the earlier one. Separate analyses were done for accidents in which the driver was reported as 'dozed at the wheel', single-vehicle accidents, and other-vehicle accidents. All analyses showed a significant effect of number of hours driving, with about twice as many accidents occurring in the second half as in the first half of the trips. The greatest deviations from the expected accident frequencies tended to come from the fifth and sixth hours on, though the difference tended to diminish or disappear during the ninth and tenth hours (and earlier for other-vehicle accidents). In a straight comparison between working times of 14 or more hours duration and those of 10 hours or less, Hamelin (1980) found that the accident risk for truck drivers was 2·5–3 times higher in the former than it was in the latter.

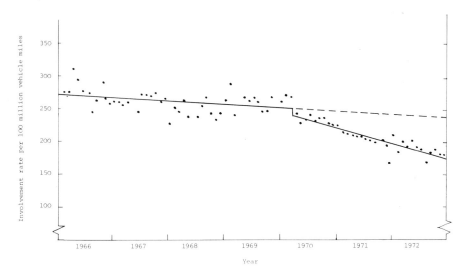

Figure 3.5. HGV (over $1\frac{1}{2}$ tons unladen weight) involvement rate in accidents per 100 million vehicle miles, Great Britain, 1966–1972. Source: Department of Transport 1981.

It is possible to speculate on the effects of legislation curtailing drivers' hours on accident rates by comparing the accident involvement rate of HGVs in the periods leading up to and following the implementation of the Transport Act (1968) in the UK (see Figure 3.5). The provisions on drivers' hours of this act, restricting the total permissible daily driving time from 11 hours to 10, came into force in spring 1970. Beginning around that time was a marked decline in the involvement rate of HGVs, which continued through 1971 and 1972, though the rate of decline has levelled off somewhat in more recent years. Unfortunately for the fatigue researcher, the Transport Act also included provisions concerning vehicle testing, overloading and driver licensing and testing, most of which were introduced at around the same period, and are thus also contenders to the claim of having been effective safety measures. The contribution of the limitation of drivers' hours to the reduction in accidents is thus impossible to evaluate.

Comparable evidence from other industrial workers may throw some light on the relation of hours of work to accidents. Early work of the Industrial Health Research Board in the UK in World War I, particularly among munitions workers, investigated the rather extreme working shifts that were then operative. Vernon (1918) summarizes some typical results as follows:

The influence of fatigue on accidents to women was strikingly shown at the Fuze Factory when the operatives were working a 12-hour day or 75 hours a week. The women's accidents were two and a half times more numerous than in the subsequent 10 hour day period, but the men's accidents showed no difference. Also the women were treated for faintness nine times more frequently than the men, and were given sal-volatile 23 times more frequently, whereas in the subsequent 10 hour period they were treated for faintness and given sal-volatile only three times more frequently. (p. 46)

This result emphasizes the importance of factors outside work on fatigue and accidents—women were far more vulnerable to the dangers of excessive hours of work because the sexual division of domestic labour meant that they carried the burden of housework as well as factory work. Comparable factors over and above their actual driving duties may be operative with many truck drivers, particularly trampers, who have frequently to find and arrange loads, seek out accommodation in strange towns, and other activities which may prolong their working day considerably. Vernon *et al.* (1931) found that age affected susceptibility to accidents among coal-miners; increasing the shift length from seven and a half to eight hours increased the accident rate of older miners much more than that of younger miners.

However, apart from these studies which have investigated the effects of changes in shift lengths, most studies of industrial accidents have found that the fluctuations in accident rate throughout the shift is much more closely related to the rate of work than to any predicted curve of fatigue—accidents come when people are working fastest (Vernon 1918, 1926, 1945, Powell *et al.* 1971, Hale and Hale 1972). It is certainly not necessary for a truck driver to have been working for very long before he falls asleep at the wheel, as the study by the Bureau of Motor Carrier Safety (1971) amply testifies: 28% of the drivers who fell asleep had driven three consecutive hours or less before the accident and 74% had driven seven hours or less.

However, drivers do tend to fall asleep at particular times of the 24-hour cycle; and predictably the early morning hours show a very high incidence of this type of accident. Over 80% of the falling asleep accidents in the Bureau of Motor Carrier Safety study (*op. cit.*) occurred in the morning hours. Prokop and Prokop (1955) found that among their sample those drivers who had fallen asleep at the wheel tended to have done so either between the hours beginning at 2300 and 0500 hrs (58%) or between the hours beginning at 1200 and 1500 hrs (25%); leaving only 17% of cases occurring in the remaining 13 hours.

Other studies of road accidents report similar patterns: Crowley and Hearne (1972) and Baker (1968) compared accident incidence rates with hourly traffic volumes; both studies identified the hours between 2200 or 2300 hrs and 0600 hrs as being particularly dangerous, the former study using all injury-producing accidents in Ireland in 1970, and the latter single-vehicle accidents on certain routes in the USA. Robinson *et al.* (1969), studying the frequency of rural injury accidents in the USA, found that tractor-trailer units were particularly heavily involved in accidents in the early morning hours in comparison to other times of the day and in comparison to other types of vehicle (though the relation of this pattern to traffic volume is now known).

Apart from drowsiness and falling asleep, there are two other factors which can be invoked to explain this higher-than-expected accident rate during the late night and early morning hours—these are darkness and alcohol usage. If darkness were the primary factor, one would expect the accident rate to be more or less constant throughout the hours of darkness; however, as Baker's (1968) data show, this is not the case. Alcohol undoubtedly makes an important contribution to accidents during these critical hours; it is, however, possible to control out this effect to some extent by considering only those accidents where the driver in question has been shown by a breathalyser test to have had less than the legal minimum concentration of alcohol.

This does not, of course, exclude the possibility of the driver of the other vehicle involved (where there is one) being drunk. Haddon (1971) quotes evidence that many crashes involving large trucks are initiated by other drivers or pedestrians who have high blood alcohol concentrations, whereas truck drivers themselves are far less likely than other drivers to be impaired due to alcohol. UK accident statistics confirm that HGV drivers have a much lower rate of being found with illegal breath alcohol concentrations per accident involvement than other drivers (TRRL 1976).

Table 3.5 shows, for three periods of the day, frequencies of different types of accidents involving HGVs in which the driver had a negative breath test result. The daily distributions of single-vehicle accidents and rear-end collisions are very similar (the differences are statistically insignificant) and together differ markedly from the distribution of all other accidents (and this difference is highly significant: $\chi^2 = 13.33$, d.f. $= 2$, $p < 0.001$); furthermore, comparing the early morning hours with all other times of the day shows a significantly greater proportion of accidents at this time to involve only one vehicle or two vehicles travelling in the same direction, compared to the rest of the day. Recalling that these two types of accident are those in which fatigue and falling asleep are most often implicated, a general pattern of accident causes begins to emerge.

Table 3.5. Accidents involving HGVs (over $4\frac{1}{2}$ tons u.w.) related to movements of the vehicles involved before the accident and time of day, Great Britain 1974 (only accidents in which a breath test was negative are included).

	Time of day						
	Midnight–07 59		08 00–17 59		18 00–23 59		
Vehicles and their movements	Number	%	Number	%	Number	%	Total
One moving vehicle involved	20	35·1	42	23·6	16	19·0	78
Two vehicles moving in same direction	15	26·3	43	24·2	10	11·9	68
All others	22	38·6	93	52·2	58	69·1	173
Total	57	100	178	100	84	100	319

Source: TRRL (1976).

Harris's (1977) study confirms this pattern of the diurnal distribution of different types of accident. He compared three types of accident (those in which the driver had 'dozed at the wheel', single-vehicle accidents, and accidents involving other vehicles) with the daily variation in the level of truck traffic. This he obtained from a random sample of 2000 driver logs. He was not, however, able to obtain data on the diurnal distribution of total traffic density. Other-vehicle accidents tended to follow the truck traffic density fairly closely. Single-vehicle accidents, on the other hand, were almost two and a half times more frequent between midnight and 0800 hrs than would be expected on the basis of the traffic distribution (46% of the accidents, 19% of the traffic). But, most dramatically, 66% of 'dozed at wheel' accidents occurred during this time. Most of these were single-vehicle accidents; those 'dozed at wheel' accidents that involved another vehicle showed a slight circadian effect with 34% occurring in these early morning hours. In a comparison of accident frequencies with truck traffic density Hamelin (1980) was able to demonstrate that the accident risk for truck drivers during the night-time hours (2000 hrs to 0600 hrs) was around twice that obtaining during the remaining daytime hours.

The industrial accident literature has produced conflicting evidence concerning relative accident rates in the night-shift and the day-shift. Hale and Hale (1972) quote evidence that in organizations where a continuous three-shift system is in operation the number of accidents in the night-shift is lower than on the day-shift, and put forward several suggestions that have been used to account for this pattern, namely at night there may be a slower work pace, greater caution, fewer distractions, different age structure of the work force, etc. On the other hand Quaas and Tunsch (1972), studying accidents in a metallurgic plant, found a higher frequency of accidents in the night-shift compared to the morning and afternoon shifts. There was a significant increase from the first to the third successive night-shift and the fourth (the last in a series of night-shifts) had only slightly less than the third. Kogi and Ohta (1975) studied near-accidents involving locomotive drivers; such incidents involving drowsiness were far more frequent during the night hours compared to other times of the day, while incidents involving unpredicted obstacles, for example, had their highest frequency during the daytime hours. On the first of two consecutive night-shifts, drowsiness was largely a problem after the driver had worked for more than 8–14 hours; however, on the second night of duty most cases of drowsing appeared in the initial 2–3 hours of duty.

One can conclude from the studies of accident frequencies involving trucks that there is a significant relationship between hours of driving and the risk of accident, though the precise form of this relationship is not clear. Very long hours of driving (in excess of 14 hours) are clearly associated with a higher accident rate, but the rate seems to fluctuate when shorter driving periods are examined (comparing eight or 10 hours with five or six, for example). Driving at night is also associated with a higher accident risk than driving during the daytime hours, and it seems that both driver 'dozed at wheel' and single-vehicle accidents are disproportionately common at night.

However, there is much that remains to be explained about the relationship between time and accident risk. One major problem with these analyses has been that they have treated time driving and time of day as separate independent factors, whereas, in fact, there are several different temporal factors, potentially related to fatigue, which interact and are confounded in these studies. Such factors include the duration of sleep periods, length of time since sleep, as well as hours of work and driving, and time of day. Pure time–of–driving effects and pure time-of-day effects never occur in reality, they both interact and must also be influenced by the timing and duration of sleep.

An interesting demonstration of the interaction between shift time and driving time is provided by Pokorny *et al.* (1981) in an analysis of accidents involving buses in one company. They showed that: (*a*) the early shift (starting time 0530 hrs to 1000 hrs) contributed a much higher proportion of the accidents (corrected for mileage exposure) than the late shift (starting time 1300 hrs to 1700 hrs); (*b*) within each shift, earlier starting times were associated with higher accident rates; and (*c*) that the two shifts differed quite markedly in the relationship between hours of driving and accident rate. Higher accident rates in the late shift were found only during the early hours of driving duty, while in the early shift peak accident rates were seen in the third and fourth hours and to a lesser extent in the eighth and final hour of the shift. They conclude that hour of the day as such does not have a substantial impact on accident risk and that there is a significant interaction between shifts and hours of driving duty.

The drivers in this sample did not work for longer than eight hours, nor between 0100 hrs and 0530 hrs. They thus cannot provide an answer to the role of fatigue in the rather more rigorous temporal conditions typical of the truck driver. However, they do suggest that the peculiar pattern of driving–time related results in the analyses of Harris and Mackie and Miller might be due to the confounding of two (or more) distinct shift related patterns, and secondly they open the possibility that if the time of day throughout which the whole of the driver's shift occurred was taken into account, rather than just the time of day of the accident, a much fuller picture of the contribution of shift work to accident occurrence would emerge. Clearly, if the comprehensive statistics of the US Bureau of Motor Carrier Safety (used by Mackie and Miller) could be adapted to this kind of analysis it might begin to be possible to weigh in realistic terms the contribution of shifts and hours of work and driving to the causation of accidents. At the very least this study does imply that early morning starting times (with the disruption of sleep with which they are usually associated) are a further factor involved in the causation of accidents which deserves attention in relation to truck safety.

3.5. *Concluding remarks*

The evidence is clear that truck driving is a dangerous profession, and that trucks are a major cause of death and injury to other road users. The evidence that fatigue and falling asleep are important factors in this pattern is not so clear-cut. Accidents involving trucks which are attributable to an impairment in the driver's functioning are overwhelmingly cases of being asleep at the wheel or being very fatigued. Fatigue has also been implicated in driver inattention accidents. Together these are only a very small proportion of all truck accidents, but the full contribution of fatigue to accidents is probably much greater than this. Fatigue and falling asleep are particularly associated with single-vehicle accidents and rear-end collisions, and these types of accident are disporportionately common at night. Single-vehicle accidents have higher rates of death to truck occupants than accidents involving other types of vehicle. Based on a comparison with exposure to risk, both long hours of work and driving at night are associated with a much higher rate of accidents than shorter hours and daytime driving. While there does not seem to be a clear linear relationship between the number of hours of driving and accident risk, the accident rate in the second half of driving trips has been found to be about twice as high as in the first half. However, pure time of day effects and pure duration of work effects never occur in isolation; they are inevitably confounded, and this may explain the confusing pattern of relationships between accident risk and time of driving. There is evidence of a relationship between shift start times and accident risk in bus drivers, and this implicates early morning shift start times in a high proportion of daytime accidents. There are likely to be other interactions between shift start times and shift durations which need to be explored before a clearer picture of the relationship between accidents and working and driving time is obtained.

Chapter 4
The Skill of Driving

4.1. Introduction

This chapter is the first of three which focus on experimental manipulations of the driver, recording aspects of his performance, his physiological responses and his feelings, in contrast to the previous two chapters which have attempted to outline the actual situation in which drivers operate.

It is appropriate to begin with a brief exposition of the nature of driving skill, and an outline of the structure of this chapter. This is followed by a consideration of some of the differences involved in driving a truck compared to a car, a discussion of some techniques and methods that have been used in driving research, and of some possible criteria for the evaluation of driving performance.

Some definitions of driving skill are notable more for their linguistic extravagance than for the information they carry. Thus, for example, Kao (1969) describes driving as 'a closed-loop, feedback-regulated, driver–vehicle–road tracking system with well organized signal, force, spatial, temporal and motion properties' and in another paper (Smith *et al.* 1970) the car is included in the picture as 'a wheeled exoskeleton of the driver'. A more practical analysis is given by Gibson and Crooks (1938), who emphasize that the primary task of the driver is to maintain a 'field of safe travel' into which his vehicle can continue moving without being impeded by obstacles or leaving the road. He must also have an idea of the minimum stopping distance of the car, which must, of course, be smaller than his field of safe travel. Successive fields of safe travel, it goes without saying, do need to add up to the driver's desired route.

In order to achieve this objective, the driver has to be able to perform a relatively complex interrelated set of skills. At the most basic level are those involved in the physical control of the vehicle—steering it along the driving lane, and maintaining adequate speed through the co-ordinated use of brake, clutch, gears and accelerator.

At the same time he has to detect and respond appropriately to any of a range of relatively discrete signals; these would include road-signs, obstacles in the roadway, and the signals and movements of other road-users. He has also to interact with other vehicles in a variety of manoeuvres such as following, overtaking and passing, lane-merging, and negotiating other vehicles at intersections. Such manoeuvres involve the driver's ability to make critical decisions on the basis of what is often very imprecise perceptual information, as well as his interpretations of other drivers' intentions. These interpretations will tend to be made according to a set of, usually implicit, assumptions about drivers' behaviour, some of which may be common to nearly all drivers, but others which may vary from driver to driver. Shor (1964) discusses some aspects of this in a preliminary way. The driver is also occasionally confronted with any of a number of emergency situations where the consequences of his action (or inaction) are potentially disastrous, and where tolerances of error are low, and opportunities for learning are limited. Such situations might include a sudden blow-out at speed, brake failure, skidding on unexpected ice patches, a child running into the road, and many others.

The sections in this chapter broadly follow the pattern of this analysis of driving skill; each component of the skill is described in some detail and its potential for deterioration under 'fatiguing conditions' is evaluated. A section on the tracking component of driving (steering and speed control) is followed by one on the driver's detection of discrete signals both inside and outside the car. Next, manoeuvres involving more than one vehicle are analysed with a particular emphasis on following and overtaking. There is no section on drivers' responses to emergency situations due to a virtually complete lack of relevant research. Instead, there is a section on the mental demands of driving and attempts to measure the information load of the task on the driver.

The most obvious difference between a truck cab and a saloon car is their respective heights above the road. The truck driver has a far superior view of the road ahead, being able to see over the top of most other vehicles and over hedges and walls; he is thus in a position to anticipate events in advance of a car driver, and the behaviour of the latter will often seem foolish and short-sighted from the truck driver's vantage point. However, the size and weight of a truck, with its poor manoeuvrability and inadequate braking performance make such anticipation a necessity, and in Gibson's terms the truck driver has to maintain a field of safe travel that extends much further than that of the car driver. Fully laden trucks also tend to be underpowered and take time to build up speed; thus part of the skill of driving is avoiding the necessity to slow down, and this is often not appreciated by the drivers of lighter and more 'flexible' vehicles. There is a greater danger of skidding in trucks than other vehicles, and articulated vehicles have the additional problem of jack-knifing. In most in-dustrialized countries, increasingly strict training requirements are stipulated for HGV drivers, requiring a level of skill far in advanced of a car driving licence. This training tends to emphasize, among other things, systematic patterns of observation, and the anticipation of road and traffic events.

Experimental investigations of driving performance vary in the amount of control the experimenter has over the situation, and the degree of realism in the situation; by and large these two goals—realism and tight experimental control—conflict with each other. Driving simulators allow the experimenter the greatest freedom to determine the experimental condition. No simulator can, by its very nature, be totally realistic, not least because the consequences of gross driver errors on a simulator are trivial compared to normal road driving. The comparability of operating a simulator to driving a road vehicle has seldom been seriously investigated. Thus much simulator research can only be said to have superficial 'face validity'. Of course, simulators do have the advantage that they permit investigation of driving situations that could not, for safety reasons, be manipulated in real life. Simulation research is also usually cheaper and faster to run than real life research.

Test tracks, often on disused airfields, or race tracks, can for some limited purposes provide a reasonable compromise between control and realism. Studies using public roads with normal traffic can vary widely in the extent to which the experimenter intervenes in the situation to eliminate unwanted effects. Such studies will vary according to the extent to which the routes are predetermined, whether the subject is carrying out his normal business, whether or not there is an observer in the cab, and the degree of instrumentation the subject has to tolerate both surrounding and attached to him. A few investigators have managed to conduct more or less naturalistic investigations in which they have not interfered at all with the normal movements of traffic. However, most fatigue research requires the investigator to be able to study a driver over fairly long periods, which usually involves his explicit co-operation. As soon as the driver is aware he is part of an experiment, this will affect his response to the situation, according to how he perceives the demand characteristics of the experiment. Subjects can often perform prodigious feats in once-off experimental situations that would be impossible if those conditions were reproduced daily in their working lives. These considerations must be borne in mind when evaluating the research to be presented in the following sections. They do not imply that the results of this research are inapplicable outside the laboratory in the 'real world', but they do imply that such experimental research, on its own, cannot be expected to provide a definitive answer to questions relating to prolonged driving, or particular driving shifts and the driver's ability to maintain a high level of skill.

The final point of discussion concerns the evaluation of driving performance—what criteria are to be used in judging whether driving is good or bad, safe or unsafe? The obvious criterion is the occurrence of accidents; however, there are various complications in this. Accidents are usually the result of a multiplicity of contributory factors, in which an identifiable failure in driving skill may only play a small part, or not at all; thus many accidents involve drivers who have committed no driving fault (by any other criterion of driving skill), and it is often hard to apportion blame among all the individual drivers involved. Conversely, accidents do not necessarily occur to drivers whose skill is severely impaired by defective vision, excessive alcohol or other factors. Although these would be expected to make accidents more likely, the point is

that wide margins of error are possible in driving and any criterion of driving skill
should reflect this. Accidents are rare occurrences in terms of the driver's exposure to
risk, their manipulation in an experimental situation is obviously unethical and their
causes are multifarious and interacting; thus it is only a very large-scale study that can
begin to point to possible relationships between patterns of driving behaviour and
accidents. Therefore, although the ultimate reference point for driving research
should reflect road safety (actual accident rates), it is necessary to postulate at an
intermediate level of analysis some criterion by which to judge driving skill.

There are some situations in driving where safety margins can be stated quite
precisely (though problems of recording and measurement are another matter). For
example, variations in time headway represent objective changes in the time available
to a following driver to respond to a sudden deceleration of the lead vehicle.
Assuming that the probability of the lead vehicle's deceleration remains constant, as
do the following driver's level of attention and speed of reaction, then the variation in
headway will bear a direct relation to safety or risk of collision. In principle it should
be possible to include in the equation variation in the other factors (which are assumed
to be constant above). Similarly with overtaking, lane-merging, and at intersections,
the gaps that drivers accept or reject can be evaluated in terms of 'safety margins',
though whether an accident actually occurs will depend on the behaviour of the other
drivers involved, and how the gap-accepting driver modifies his behaviour as the
situation develops.

With other aspects of driving it is not so easy to specify safety margins. In such
cases the judgement of some 'expert' practioner of the skill is often employed as an
evaluation criterion. This judgement in turn may or may not reflect explicit criteria
held by the judge as to what constitutes driving skill. A particularly sophisticated
example of this type of approach is the series of investigations by Quenault (1967 a, b,
1968 a, b), though this has been directed more at identifying stable individual
differences in driving style rather than changing levels of skill within one individual.
Driving is a very complex skill and Quenault used up to three observers recording
from one driver at any one time. Such intensive study is not quite so easy over
prolonged periods of driving and one observer would have to be selective about the
variables he recorded; the evidence validating any one variable or set of variables as
being crucial to safety does not exist. Also, this method throws up the problems of the
reliability of the observer between different occasions, or over long periods, and the
effects of the presence of the observer on the behaviour of the driver.

Experimental studies which are designed around the automatic recording of
driver behaviour have tended to rely on a criterion of consistency by which to judge
skill. Typically it is accelerator, brake, steering, or other control movements which
are recorded in these studies. One of the earlier proponents of the view that a
consistent pattern of control movements denotes a high level of skill was Lewis (1956).
There is probably some truth in this view, in that consistency is part of the everyday
concept of skill. However, measuring the consistency of control movements depends
on having close control of the stimuli to which they are responses—relatively

straightforward on a test track, but much more difficult on the open road. Without being able to specify the stimuli controlling the vehicle control actions, the concept of consistency becomes meaningless; and even with such detailed knowledge the significance of any change in pattern of response may be trivial in terms of safety.

Other researchers, particularly in this field of driving fatigue, have attempted to approach the question of driving safety by specifying changes in the driver's status or capacity. Typically this is done by recording his performance on a subsidiary task executed concurrently with driving. The rationale for such studies may be that one can measure the amount of 'spare mental capacity'. Alternatively, such a subsidiary task may provide some kind of analogue to aspects of driving which are difficult to control—for example a subsidiary vigilance task could be said to represent the detection of important signals from the roadside or within the car. There are two major drawbacks to this approach, however: the introduction of a novel additional task must inevitably affect the main task of driving; and the theoretical basis of this research in the single-channel hypothesis of mental capacity and the theory of signal detection, apart from being rather tentative and controversial, is probably not appropriately applied to the special circumstances of the driving situation.

Unfortunately, there have been few attempts to validate these various experimental methodologies and conceptions of skill against their ability to predict the occurrence of accidents under normal driving conditions, and the results of these few have often been negative. Tarrants (1960) found no relationship between drivers' accident records and their performance on a driving simulator. Edwards *et al.* (1969) found no relationship between scores of driving performance recorded on a simulator and a car on the road, and very low relationships between accident and traffic violation histories and all performance measures. Two small studies provide slightly more optimistic results: Babarik (1966) reports that measures of reaction time correlate with the incidence of rear-end collisions among New York taxi drivers; and Henderson (1977) reports that among Northern Ireland truck and bus drivers, patterns of mirror usage bear some relationship to their accident history. Although rudimentary, it is studies like these which may form the beginnings of a more detailed picture of the relationship between accidents of particular types and the antecedent pattern of skill degradation which led up to them.

Although, as has been said before, the evidence discussed in the following pages cannot be expected to give definitive answers to questions relating to the consequences of particular practices among drivers, the potential is there to suggest ways in which the quality of driving skill may be susceptible to deterioration under stressful conditions.

4.2. *Driving as a dual-tracking task*

The two most basic requirements of driving are to keep the vehicle within the driving lane at the same time as maintaining adequate speed. These two aspects of the task are, of course, closely interrelated—desired speed will reflect, among other things, the contour characteristics of the road; and the difficulty of the steering task will depend on the speed of travel. This much is obvious, but when we try to build up a detailed picture of the complex of sensory cues to which the driver is responding in monitoring his speed and direction and attempt to relate this to the driver's pattern of control movements, we find, surprisingly, that only a partial and incomplete picture is possible, and this is particularly true when we want to evaluate the driver's performance along some kind of 'fatigue' dimension. We will start with the sensing of motion, direction of motion and position, and then consider how these stimulus aspects might relate to the driver's control actions.

At very slow rates of motion, movement has to be inferred from changes in position; at higher rates motion can be perceived directly (compare the speeds of the minute and second hands of a watch); while at still higher speeds motion appears as a blur. When we as observers are moving through our surroundings the apparent motion of objects in relation to us is proportional to their real distance from us (so long as we are going in a straight line). Thus when driving through open country the horizon may appear not to be moving until we have noticed it to have changed; an object in the middle distance will move steadily and perceptibly across our field of vision, while the road verge will rush past in a blur. (This is the basis of the cue to distance known as 'motion parallax'). The apparent movement of our surroundings relative to us follows a pattern well described by J. J. Gibson (1950, 1954, Gibson *et al.* 1955). As we move across a flat terrain, objects in our field of view follow certain lines of flow. These lines of flow radiate out from the one static point in our visual field (other than the car) which is the point towards which we are heading—the 'focus of expansion' (see Figure 4.1). The faster we are going, obviously the greater the proportion of the visual field that is blurred, and only at greater distances ahead can we fixate (Connolly 1966), though this receding point of concentration also has the effect

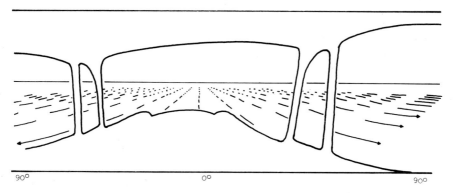

90°   0°   90°

Figure 4.1.   Velocity vectors on a flat terrain illustrating the focus of expansion. If the page is curved in a semicircle and one eye is placed as near to the centre as it can focus, a better visual representation is achieved. Adapted from Michaels (1973).

of maintaining a more or less constant preview time of the road ahead. When we face the direction of travel, the greatest angular velocity in our field of view will be in the peripheral rather than the frontal areas, and Salvatore's (1968) findings do indeed suggest that the peripheral visual field does provide more accurate velocity information than the frontal field.

Following further investigation, Salvatore (1969) concluded that peripheral visual information is a more reliable cue to velocity than either auditory or kinaesthetic–vestibular cues. Evans (1970), on the other hand, found that eliminating auditory cues distorted speed estimation to a greater extent than eliminating visual cues; subjects deprived of auditory cues from engine and road noise tended to underestimate speed. It seems likely that both auditory and peripheral visual cues are important in accurate speed estimation; furthermore, in most situations drivers will tend to rely upon these cues rather than the direct evidence of the speedometer, which is sampled only rarely (Denton 1968).

While kinaesthetic and vestibular cues are not important in providing information about absolute levels of speed, they are particularly well adapted to registering changes in velocity—both changes in the fore and aft axis due to accelerator and brake usage, and lateral accelerations involved in cornering and skidding, for example. Steady states of motion produce a response from pressure receptors under the skin, and from the hair cells distorted by the otolith in the vestibular apparatus, that is very similar to the response in a stationary state.

Judgement of the direction in which the vehicle is heading probably utilizes two more or less distinct cues. Gibson (1950) suggests that steering a car is "less a matter of aligning the car with the road than it is a matter of keeping the focus of expansion in the direction one must go" (p. 128). Gordon and Michaels (1965) contradict this view, claiming that it is often impossible for drivers to locate the focus of expansion, particularly on curved or hilly roads, and that the primary cues to the heading of the vehicle are the borders and edgemarkings of the road. Both of these viewpoints get some support from findings reported by Rockwell (1972 a) using a corneal-reflection technique to photographically record driver's eye-fixations. Instructions emphasizing lateral control accuracy led to a greater concentration of fixations near the focus of expansion, suggesting that general directional information is obtained foveally from around the focus of expansion. However, it also seems likely that the lateral placement of the vehicle within the driving lane is largely controlled through peripheral visual monitoring, with the occasional foveal fixation on lane markers and other vehicles, when the situation demands this (Mourant *et al.* 1969). This view is supported by Shinar *et al.* (1977), who suggest that along straight roads the driver tends to look at the focus of expansion and adjust to changes in road position detected peripherally, whereas on curves, following a preview scanning of the curve, there is an alternation of fixations of the road ahead (to gain directional information) and the road edge (to gain positional information).

The distinction between the monitoring of the vehicle's lane position and its direction of travel is given further support by Gordon (1966); he occluded the driver's peripheral vision and interpreted the resulting pattern of foveal fixations as reflecting a trade-off between fixations immediately ahead and to the side giving positional information and fixations further ahead of the car giving general directional information. The general importance of peripheral visual information to the driver is emphasized by Mortimer (1967); with their field of view blacked out and replaced by a compensatory tracking task (centring a visual 'blip' on a CRT display), the drivers in this study would not exceed 15 m.p.h. and felt disorientated; however, when allowed additional peripheral information their maximum speed reached 25 m.p.h.!

But the role of the focus of expansion in vehicle guidance has been questioned by Riemersma (1982a and b), who suggests that the optical cues of lateral position, lateral speed and heading rate are the effective cues in controlling the vehicle's path and that heading direction is not controlled directly.

There is some evidence which suggests that prolonged driving and deprivation of sleep is associated with a change in the pattern of eye movements which might in turn imply a degradation of visual information acquisition. Kaluger and Smith's (1970) subjects wore an eye-movement camera for two separate days of driving—one day after a normal night's sleep and the other after a night with no sleep; each day's drive consisted of three 150-mile laps ($3-3\frac{1}{2}$ hours driving each)—a fairly severe test. As driving proceeded throughout the day, there was a tendency for the subjects to fixate progressively closer to the car; thus, in terms of safety margins, preview times of the road ahead which at the beginning of the day averaged around five seconds, at the end averaged two seconds; rest breaks were accompanied by some recovery, however. As well as this, there was a gradual shift to the right of the 'mean locus of fixation', away from the focus of expansion. This trend was true only for the day's driving following a night's sleep; in the sleep-deprived condition the 'mean locus of fixation' started well to the right compared to the beginning of the day after sleep (see Figure 4.2). Corresponding to these trends was a tendency for less viewing time to be spent close to the focus of expansion, and for eye movements to be less concentrated in a small area. Again, rest breaks were associated with a 'recovery' towards the pattern at the beginning of the post-sleep day. Kaluger and Smith interpret their results as reflecting a degradation in peripheral visual sensitivity and the compensation for this by the allocation of more foveal viewing time to areas normally monitored by the peripheral retina. Thus the downward and rightward trend of eye movements could represent an attempt to maximize velocity information. (The study was conducted in the USA with, of course, driving on the right-hand side of the road.) Confirmation of the detrimental effect of sleep-loss on the processing of information from the peripheral visual field comes from laboratory experiments by Sanders and Reitsma (1982a and b).

These results are very interesting, but suggestive rather than conclusive. The sort of corroborating evidence that would be particularly useful would be some kind of demonstration that under conditions similar to those of Kaluger and Smith's study, drivers were indeed less capable of maintaining an appropriate, or desired, speed, or

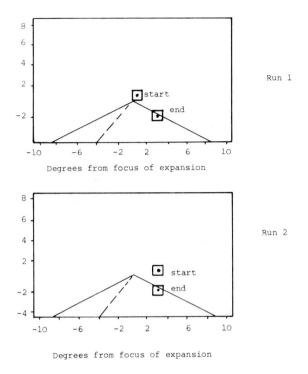

Figure 4.2.   Approximate positions of the mean locus of fixation at the end of run 1 (after sleep) and run 2 (no sleep). After Kaluger and Smith (1972).

that their control of their vehicle's position or direction of travel had suffered in some way.

  Much of the experimental research on the driver's control of the vehicle which might provide this evidence has broadly been based on a set of postulates set out by Platt (1964), which read as follows:

  1. Driver fatigue will have effect on steering wheel reversal rates, speed change rates, and average speed of the vehicle.

  2. As the driver becomes fatigued he will accept wider tolerances of both vehicle tracking and speed control.

  3. As a driver gets tired his speed may increase or decrease depending on whether his sensitivity to speed change or steering reversal rate is lost first.

  4. The driver will usually take more risks as he becomes more fatigued. This will be indicated as an increase in tracking tolerance, and consequently, a decrease in steering reversals if speed is constant. Degree of risk may also be indicated by an increase in the speed of the vehicle.

  5. As the driver becomes tired his speed change rate increases but he usually makes some effort to keep it within balance by accelerator reversals.

6. The most severe fatigue is encountered in driving when the speed change rate increases and accelerator reversal rate decreases. This indicates the driver has ceased to care about speed control. (p. 356)

From this set of assumptions Platt derived a number of mathematical equations allowing him to derive one overall score of the amount of driving deterioration. This procedure did rather beg the question as to whether the postulated changes (which were presumably only based on his intuitive reading of the situation) were in fact all useful indicants of the 'fatigue' process. Subsequent to Platt's (1964) study, investigators have reported results from each variable measured separately, while in Platt's analysis virtually any change from initial levels contributes to the ultimate 'deterioration' score.

There are two major difficulties in accepting Platt's formulation as it stands: firstly, the relationship between measures of control movements (such as steering and accelerator reversals) and such concepts as 'tracking tolerance' and 'speed control' is considerably more complicated than the model suggests; and secondly, it violates the second of Muscio's (1921) conditions for the adequacy of any fatigue test, which states: 'that we have some method *other than the use of a suggested fatigue test* by which we can *know* that different degrees of fatigue are present at different times' (emphasis in original). Thus if we have a driver who is showing a declining number of reversals of steering wheel movement per unit time, we cannot tell whether this represents a decreasing 'tracking tolerance' level or some other factor to do with road or traffic conditions; and even if it does represent 'declining tracking tolerance', that this is due to 'fatigue' in any sense is not at all clear.

One of the most difficult measures to interpret has indeed been steering-wheel reversal rate, which has been one of the most common measures in 'driving fatigue' studies. It is possible to give only a very speculative account of the determinants of steering wheel activity. McLean and Hoffman (1971) put forward two tentative explanations of the results of a frequency spectrum analysis of steering wheel movements. They tested two drivers over straight and curved tracks, and the analysis revealed two peaks of steering activity—a dominant peak between roughly 0·1 and 0·3 Hz and a smaller peak between 0·35 and 0·6 Hz. One suggestion is that the dominant low-frequency peak represents general directional control movements based on the driver's preview of the road ahead, with the smaller peak reflecting occasional higher frequency adjustments to his lateral position on the road. Drivers do tend to maintain a preview time of between 2 and 7 seconds, and this would accord well with this suggestion. Alternatively, the low-frequency peak might relate to the control of lateral error—assuming that the driver will only adjust the lateral displacement of the vehicle when it runs out of a zone within which he either cannot perceive, or is not concerned about, changes in lateral position. The high-frequency peak would then reflect the driver's attempts to minimize the angular movement of the car (and thus is more pronounced during cornering and overtaking). These two alternative explanations are by no means mutually exclusive—some combination of them might be closest to the truth. As well as the frequency of steering movements, the amplitude of such movements is obviously important; Brown (1976 personal communication) considers that the aspect of this which is most likely to show the effects of 'fatigue' is the number of small 'hunting movements' the driver makes just within the free play of the steering wheel; these movements he suggests are related to the driver's continuous monitoring of the vehicle's heading (indeed, he has found that individual differences in steering reversal rate correlated with differences in accuracy in judging stationary vehicle headings). Again, this hypothesis, though plausible, lacks comprehensive experimental support. In a more recent paper Macdonald and Hoffman (1980) sought to show that the level of steering activity was related to the complexity of the driving task, with both high and low levels of task demand being associated with low rates of steering reversals.

The outline results of a number of studies which have recorded some aspect of the tracking components of driving under 'fatiguing' conditions are contained in Table 4.1 below.

While the five simulation studies unanimously report declining tracking accuracy as a function of time, this consistency has not been reproduced by studies on the road. Human Factors Research, Inc. (HFR) have produced the greatest amount of research on steering accuracy (nos. 14–18 in Table 4.1). These studies used an observer in the cab to score the number of lane drift incidents, and their more recent study (Mackie and Miller 1978) employed an electronic device for sensing edge- and centre-markings on the road. This device, which has been described by O'Hanlon (1976), is similar to that used by Riemersma *et al.* (1976).

Lane drifts are a rare phenomenon, at least among the professional drivers who were the subjects in these studies. They are most likely to occur during the late night and early morning hours, at which times there also tends to be the greatest incidence of other symptoms of drowsiness in the driver. Mackie and Miller also noticed that the frequency of drifts during night driving was rather less for 'sleeper drivers' when they had to perform a loading task as well as their normal driving duties; though this was not the case for relay drivers on rotating shifts. This moderate amount of physical activity thus sometimes appears to counteract lane drifting (though not necessarily other expressions of drowsiness). All the drivers were relatively unused to performing such work. There are also indications from these studies of a greater tendency for lane drifts to occur after long periods of driving and towards the end of a week of prolonged driving, though this is by no means always the case. Results using the lane tracking device are also very variable, but within this variability the pattern is broadly the same, with there being a tendency for a falling off in tracking accuracy to occur more often in the last hour or so of driving, towards the end of a week of driving, or occasionally during late night and early morning shifts, than at other times. Thus significant decrements in tracking accuracy were observed as early as the fifth to sixth hour of driving during the shorter irregular shifts (6 hours driving in total), and as early as the seventh to eighth hours of 9-hour driving periods, particularly when these occurred towards the end of the working week. The great variability in results needs

Table 4.1. Main findings of studies employing measures of speed and steering performance over prolonged driving periods.

Investigator(s)	Type of vehicle	Length of drive (h)	Lane drift	Steering wheel reversals	Speed	Accelerator reversals and brake applications
1. Mast *et al.* (1966)	Simulator	6	Tracking accuracy declined		Maintenance of prescribed speed deteriorated particularly after 4th hour	
2. Heimstra (1970)	Simulator	6	Increase in tracking error (particularly in groups receiving contingent shocks)		Increased deviation from prescribed speed only in shock group	
3. Dureman and Boden (1972)	Simulator	4	Increase in tracking error (less so for contingent shock group)			
4. Sussman and Morris (1970)	Simulator	4	Increase in tracking error (There was a significant negative correlation between steering reversal rate and tracking error of −0·54)	Significant decrease in fine (2°) reversals	No time effects on velocity control	
5. Snook and Dolliver (1976)	Simulator	3	Increase in tracking error	Slight decrease only in second half of run (overall not significant)	Increase in speed variation	
6. McFarland and Mosely (1954)	Bus	$3\frac{1}{2}$		Steering activity lower at end than beginning of run, but overall trend not significant		

7. Shaw (1957)	Car	6			No time effects on accelerator movements
8. Platt (1964)	Car	Sunday—Tuesday 2 drivers alternating every 1½ h approx	'Overall driving performance' declined		
9. Greenshields (1966)	Car	6	Very little variation in speed	Some drivers showed decline with time, others not	No consistent variation in either accelerator or brake movements
10. Safford and Rockwell (1967)	Car	24 h–5 subjects 21 h–2 subjects	Velocity variance increased, velocity mean decreased	Increase in steering reversals with time	Accelerator reversals increased
11. Brown (1967a)	Car	12	No significant changes in longitudinal accelerations	No significant changes (very slight decrease)	
12. Lauer and Suhr (1959)	Car	3–4	No significant changes in mean or variance	'Lateral placement' accuracy declined in 'no pause' condition	
13. Michaut and Pottier (1964)	Car (closed track)	3	Slight decrease in mean speed (not significant)	Significant decrease in steering corrections from 1st–6th half hour. Small amplitude movements (1° 38') *increased*. Large amplitude movements (4° 54') *decreased*	3 subjects out of 11 showed a decreasing rate of accelerator movements
14. O'Hanlon (1971)	Van		Sketchy presentation of results—no evidence for any trends		

Fatigue, safety and the truck driver

Table 4.1. (cont.).

Investigator(s)	Type of vehicle	Length of drive (h)	Lane drift	Steering wheel reversals	Speed	Accelerator reversals and brake applications
15. Harris *et al.* (1972)	HGV	≤ 10	Generally an increasing trend in frequency of lane drifts but not always	Generally downward trend in reversal rate. More pronounced after poor sleep and for older drivers. Sometimes a recovery after rest breaks		
			Trends in lane drift often tend to mimic trends in steering reversal rate but not universally			
16. O'Hanlon and Kelley (1974)	Van	5	Drifting across lane boundaries tended to increase towards end of experimental sessions	Some evidence for decline in small reversals and increases in large reversals in some experiments	Speed variability tended to increase	
17. Mackie *et al.* (1974)	Car, HGV	≤ 10	In one experiment higher rates of lane drift found in hot condition and a slight increase with time. However this was not found in subsequent experiment	Large steering reversals (over 10°) were more frequent in hot than comfortable condition, and tended to increase with time on the road	Most changes in speed a function of traffic density, but a slight suggestion that speed variance increased with time	Neither no. of brake applications nor standard deviation of accelerator motion showed any useful results

	Vehicle	Shift			
18. Mackie and Miller (1978)	HGV, bus	Up to 10 h regular, irregular and sleeper shifts	Lane tracking: variable results—performance decrements during driving shifts at end of a week of prolonged driving and in early morning hours. Lane drift: incidence most frequent in late night/early morning shifts	General tendency for fine movements ($<2°$) to decrease and coarse movements ($>2°$) to increase. Recovery with rest breaks. Effects of time of day, workload and cumulative fatigue.	
19. Sugarman and Cozad (1972)	Car	4		Fine reversals ($2°$) decreased, gross reversals ($12°+$) increased	
20. Riemersma *et al.* (1976)	Car	8	Chosen lane position did not change significantly. Standard deviation of lane position increased as a function of time. This was due to increased size of lane drifts—not greater frequency	Indices describing steering behaviour showed no systematic changes	Significantly less consistent maintenance of speed in second half of run despite lower traffic density

to be emphasized—for example, on regular daytime bus runs there was an improvement in steering accuracy during driving shifts early in the week (Mackie and Miller 1978). Riemersma et al.'s results raise questions of scoring and interpretation of tracking accuracy which would bear clarification in further investigation; they suggest that the declining tracking accuracy found during prolonged driving is a result of an increasing magnitude of lane drifts, rather than drivers tending to drift more often.

Some clarification of the relationship between simulation studies and 'on the road' studies of steering accuracy is provided by results from a 'critical tracking task' that Mackie and Miller had their drivers do before and after driving periods. Results from this portable laboratory type task more closely mirror the findings of the simulation studies than those in which actual driving performance was measured. Thus there was a fairly consistent tendency for tracking accuracy to decline after prolonged driving, to improve after rest breaks, and to be worse during late night and early morning shifts; though the addition of a loading task on top of driving duties had no effect, and bus drivers showed much less marked decrements in performance than truck drivers (whose schedules were more arduous). Results in this tracking task were in reasonably close correspondence to variations in a subjective index of fatigue. It seems that there are considerable sources of resistance within the normal driving situation to a simple linear decrement in performance with time driving, and that the process of fatigue, and the relationship between its subjective and behavioural aspects, is perhaps rather more complex in the real world than simple laboratory models would suggest.

Steering wheel reversal rate has often seemed to be the 'great white hope' of studies requiring any behavioural measure susceptible to 'fatigue effects', ever since McFarland and Mosely (1954) reported that a long-distance bus driver showed a lower rate of steering wheel activity in the final segment of the journey compared to the first. However, McFarland and Mosely's results were statistically insignificant when considered as a trend throughout the trip; and when we consider the general pattern of results from these studies, bearing in mind how extensively steering wheel measures have been used, our conclusions must be slightly disappointing. Some studies have simply reported frequency counts of all reversal of steering wheel movements irrespective of their magnitude; others have given separate analyses for small and large reversals. According to Platt's (1964) theory, fine correcting movements should decline in frequency as 'tracking tolerance' increases, and concomitantly the number of less frequent larger amplitude control movements should increase.

Results in relation to these hypotheses have been contradictory. Two studies from Calspan Corp. (nos. 4 and 19 in Table 4.1) have reported statistically significant results in support of these hypotheses. Mackie and Miller's study has produced the most comprehensive evidence on the question, involving a very close approximation to normal working conditions for their professional truck and bus driver subjects, and comprising regular daytime runs, irregular rotating shifts (spanning the full 24-hour diurnal cycle) and 'sleeper' operations in which two drivers alternate driving and sleeping en route in the cab. In the majority of cases the trends in steering-wheel reversal rate were for fine reversals to decrease in frequency and for coarse reversals to increase when comparing driving over the same road segments at the beginning and end of trips, and many of these changes were statistically significant. Other patterns of change in steering also occurred, including the opposite of the above, an increase in both fine and coarse reversals, a decrease in fine reversals but not course, and no change in pattern. However, these latter patterns were more likely to happen earlier rather than later in driving periods, at the beginning rather than at the end of the working week (suggesting a cumulative effect of prolonged work over several days), and during evening shifts (when the circadian influence might tend to counteract fatigue) rather than in late night or early morning shifts. These results are interpreted as strong evidence for a decline in steering precision due to fatigue produced by prolonged driving, and exacerbated by previous days of prolonged driving and irregular schedules rotating throughout the 24-hour cycle; though not apparently affected by having a moderately heavy loading task to perform. These effects became apparent after as little as 4–5 hours in the shorter irregular shifts, and as early as the seventh to eighth hour of regular daytime runs—essentially they seem to be a phenomenon of the last hour or two of driving.

Though the majority of the rest of the studies of steering activity have been inconclusive, two in particular provide evidence that directly contradicts the hypotheses: Safford and Rockwell (1967) found reliable increases in steering reversal rates over very long periods of driving, and Michaut and Pottier (1964) found a pattern of increasing small amplitude and decreasing large amplitude movements over time. These discrepant findings may be due, in the former case, to a failure to differentiate different magnitudes of steering movements, and, in the latter, to the relatively short driving period. Thus the balance of evidence does seem to favour Platt's hypotheses.

However, there are major problems in the interpretation of the steering wheel reversal measure—what does it actually mean in terms of driving effectiveness and safety? It seems almost too obvious to state that the steering activity of drivers is largely a function of the contours of the road and the steering characteristics of the vehicle. Yet Mackie and Miller are alone in having tried to control for the former by comparing performance over the same stretch of roadway on outward and return parts of the trip. In none of the studies is it possible to know what kind of change in tracking performance is associated with the change in frequency of steering corrections. In Mackie and Miller's study, for example, while on many occasions trends in steering reversals seem to have their counterpart in changes in tracking accuracy, they frequently do not, and occasionally they are in the opposite direction. Mackie and Miller tend to suggest that steering reversals are a more sensitive reflection of fatigue than tracking performance, but whether this is true or not it leaves unanswered the question of the relationship between steering behaviour and control over the vehicle, a question which is highlighted by the failure of Riemersma *et al.* (1976) to find any relation between the increasing trend to drift slowly away and a decline in steering activity.

There is rather less to be said about the speed–tracking aspect of driving. A number of studies have found that, as time progresses, variation in speed increases (notably nos. 5, 10, 16, 17 and 20 in Table 4.1); however, other studies have not obtained this result (4, 12 and 13). The former result is quite congruent both with Platt's scheme, and with Kaluger and Smith's (1970) hypothesis of declining peripheral visual sensitivity to velocity cues. Mean speed declined throughout the driving sessions of Safford and Rockwell's subjects; Michaut and Pottier's drivers showed a trend in the same direction, though statistically non-significant; and Platt's model predicts either an increase or decrease in speed as a consequence of 'fatigue'—however, the majority of studies have found no reliable changes in average speed. It is interesting that those changes in mean speed that have been reported have indicated a decrease; this suggests that drivers may be compensating for whatever deterioration in visual monitoring or central processing that may occur by slowing the task down, rather than increasing their speed due to visual adaptation effects as suggested by Kaluger and Smith (1970), or to increasing their risk-taking level as suggested by Platt (1964). In interpreting studies of speed and speed change it should be borne in mind that one of the major determinants of these variables is other traffic. Mackie *et al.* (1974) are almost alone in drawing attention to this aspect of their results, and it is almost certainly a significant factor in any study on the open road.

It is quite apparent from Table 4.1 that the recording of brake and accelerator usage has not produced any useful or interpretable findings.

Although the evidence is rather equivocal that prolonged driving can lead to less skilful performance of driving down the open road, there is some evidence that certain complex manoeuvring skills may suffer under such conditions. Herbert and Jaynes (1964) tested groups of army truck drivers on a battery of nine performance tests before and after different periods of driving of up to nine hours. Herbert (1963) provides details of the tests: they involved the drivers in a range of tasks from parking and reversing in a confined space, to precision steering relying only on proprioceptive feedback (their vision was obscured). Differences in performance between tests before and after driving were compared with the length of the intervening driving period; longer periods of driving were associated with a greater decrement in performance in all the tests. This evidence might be taken to suggest that whether or not the driver can continue driving down the 'open road' almost indefinitely without his control skills showing any decrement, his performance when he has to manoeuvre his truck in the confined space of a depot at his destination may be a lot worse after a long day of driving.

In conclusion, there is some evidence that drivers' peripheral visual sensitivity may be impaired both after prolonged driving, and after a night without sleep. This may affect his perception of velocity, and some findings of an increase in velocity variance under such conditions would tend to support this. The driver's sensitivity to small changes in his lateral position on the road may also be impaired, but it is unlikely that there will be interference with the acquisition of directional information, except in so far as less foveal viewing time will be spent near the focus of expansion. There is also evidence to suggest a greater likelihood (though certainly not an inevitability) of a falling off of tracking accuracy after prolonged driving, particularly following a number of days of prolonged driving, or when working irregular rotating shifts. Lane drifting, although a rare phenomenon, is disproportionately common during night driving, and drivers have been known to leave the road even in the rather artificial and unusual situation of a driving experiment. The balance of evidence seems to suggest that there might be a decline in 'steering precision' (fewer fine and more coarse steering movements) under the same sets of conditions as threaten tracking accuracy, though the relationship between steering movements, actual vehicle control and the danger of losing control of the vehicle is not at all clear.

The following account from the author's research notes illustrates well the type of 'critical incident' that seems very pertinent to this problem.

I was an observer in a truck during a night shift. At around 3.30 am the driver seemed to become very drowsy—he would yawn frequently, stretch, rub his face and neck, alternately slumping across the steering wheel and then just barely resting one hand on it; he would start nodding and then rapidly shake his head as if to wake himself up. During this time the speed of the truck was only about 15 to 20 mph (compared to the normal 40–50 mph). The driver would allow his truck to drift towards the side of the road, correct its movement in a rather gross fashion, and correct again in the opposite direction as it approached the centre lane of the road, and so on. All this was in great contrast to his normal impressive accuracy in steering. Thus the driver was slowly weaving his way down the road; and he would take corners very slowly indeed. After about twenty minutes he seemed to 'wake up' and his driving returned to normal, though he still looked tired. He later commented that the whole period of driving was a complete blank to him though he knew he had been very drowsy and 'nodded off'.

From one point of view the driver was performing quite adequately, considering that conditions were good and there was no other traffic. The steering inaccuracy was compensated for by the speed reduction, and the driver did not appear to be about to leave the road. On the other hand, it does not seem to be an ideal way of driving, and this account does focus attention on several pertinent questions, namely:

1. How typical is this pattern of behaviour?

2. What is the relationship between this pattern and 'traditional' measures of steering reversals?

3. Would such measures detect the occurrence of such a pattern, and if not, what changes in driving patterns are measured by steering reversal counts?

4. Is there any relationship between this type of driving pattern and actually leaving the road? It is by no means clear that there is.

5. What circumstances, or perhaps driver characteristics, combine to make this type of driving sequence more likely to occur?

Until research is intelligently directed at questions such as these it will only be possible to give a vague and inadequate account of the deterioration and breakdown of the skill of keeping a vehicle moving along a road.

4.3. The vigilant driver

Superimposed upon the task of guiding the vehicle down the highway is the necessity for the driver to detect and respond to a range of relatively discrete events, including traffic signs, other vehicles, pedestrians, animals and objects on the road. It is within this context that we can best interpret a number of studies that have required the driver to do a vigilance-type task while driving. The classic model for vigilance research is the monitoring of radar screens, or general military watchkeeping, neither of which are quite the same as driving, where the detection of discrete signals is only part of the overall skill sequence. There is a considerable amount of research from studies of vigilance which indicates a decrement in performance associated with time on task. This is commonly interpreted as reflecting a change in the subject's criterion of response (according to signal detection theory), whereby he is less likely to give a positive response (either true or false) as time proceeds; this being possibly due to a number of factors ranging from sensory habituation to lowered arousal level (Mackworth 1969, 1970, Broadbent 1971). According to McCormack's (1967) theoretical review, factors which can prevent this decline in performance are, in particular, knowledge of results, rest periods, unexpected extraneous stimuli and a very predictable sequence of signals.

The investigations discussed in this section all have one feature in common—they involve the presentation of a particular discrete stimulus and the recording of the response. The response parameter is normally either the time taken to respond, or the frequency and accuracy of detections. According to Buck (1966) both types of measure follow the same pattern—as detection rate decreases, reaction time increases (though when signal intensity is high or duration long, an increasing reaction time may not be matched by a fall in detection rate). Vigilance-type tasks can vary in a large number of ways, according to, for example, the frequency, duration and intensity of signals, the sensory modality through which they are received, the nature of the required response, and whether the subject is performing other tasks at the same time. While all require that the subject directs his attention towards the stimulus, senses it and responds to it (see Jerison 1967 a), different tasks put different emphasis on these three sequential functions. Traditional reaction time tests where the subject is poised with his finger over the response button staring at an about-to-be-illuminated bulb virtually eliminate variation due to the direction of attention (unless the subject is momentarily distracted); with the signal at well above sensory threshold level and the response requiring negligible skill, what is measured is speed of 'sensory–motor co-ordination'. Vigilance studies, on the other hand, place their emphasis on the detection process—making a judgement that there is a signal there. The theory of signal detection explains the typical falling detection rate as being due to the subject's increasing 'conservatism' in deciding whether there is a signal there or not. Jerison (1967 b) argues that this theory is only appropriately applied to decisions about information that has been received, and that in most vigilance experiments, where the signal is well above threshold, the reduced detection rate during prolonged continuous monitoring is better attributed to failures of attention (including daydreaming) than changes in criterion. This argument is particularly compelling when the vigilance task in question is performed concurrently with and subsidiary to driving a car, truck or simulator. After reviewing studies which have focused on the question of detection of signals, and discussing their relevance to the visual attention processes of the driver, a few studies which emphasize the speed of response will be discussed.

Two studies from the 1960s carried out under apparently very rigorous conditions of prolonged driving cast very serious doubts on the conventional wisdom that the driver's level of vigilance must suffer under such conditions. The subjects in the study by Dobbins *et al.* (1963) were army truck drivers. The experiment comprised a nine-hour working shift, of which seven hours were actual driving time, split up by six short rest periods. The drivers were asked to respond, with a foot switch, to six critical red light signals in a display mounted on the dashboard (and which also contained nine non-critical white lights). Critical signals averaged 30 per hour. Accuracy of detection was at a high level throughout the driving periods with an average rate of 83% of critical signals detected and 4% false detections. There was no performance decrement between driving periods. It was not possible to test for within-period performance changes. Brown *et al.* (1967) mounted a small light on each of the three rear-view mirrors on a car; they were illuminated simultaneously and the time the subject took to respond to them was recorded. The study was conducted during 12 hours of driving over main roads, though at fairly regular intervals throughout the day there were six test sessions on a city circuit. Figure 4.3 shows the considerable and statistically significant decrease in time taken to respond as the sessions progressed. Brown *et al.* (1970) report substantially similar results from a further study.

A series of simulator studies have similarly broadly failed to find the predicted performance deterioration associated with prolonged task performance. Mast and Heimstra (1964) compared the effects of four hours' simulated driving on a subsequent two-hour vigilance test, with the effects of a four-hour mental task and a control (no task) condition. The driving session, unlike the mental task, had no significant effect on the vigilance tests. Heimstra (1970) in a six-hour simulation experiment recorded his subjects' performance on three subsidiary tasks; they had to detect deflections of pointer in a meter dial, detect increases in the intensity of two small lights just above the simulator screen or window, and respond to a red light by depressing a foot switch. Three groups of subjects received either small electric shocks contingent upon driving errors, non-contingent shocks, or no shocks. The only significant performance change as a function of time periods was an increase in response time to the red light in the contingent shock group. The contingent shock group also had a higher error rate in the other two vigilance tasks in the fourth to sixth hours compared to the other groups; however, there was no significant effect of trials. Ellingstad and Heimstra (1970) again recorded detections of meter deflections, as well as response times of the hand and foot to two red lights. In this study however their sessions lasted for fifteen hours (three five-hour periods). After about eight hours with very little change, the rate of detection of meter movement began to improve. This improvement lasted until the eleventh hour and was followed by an increasing error rate until the fifteenth and final hour. The pedal response time improved over the 15 hours but the manual response showed no consistent change. Sussman and Morris (1970) found no significant change in response time to a light coming on near the focus of expansion on a simulator during four hours on task. And Boadle (1976), recording the detection rate of a light on the dashboard, reported a decrease in false responses but no change in sure detections over two hours; a result more consistent with a learning or practice interpretation than 'fatigue'.

There are, however, three studies which can profitably be cited to draw a marked contrast to the above results. Riemersma *et al.* (1976) had their subjects drive a car for eight hours throughout the night and while doing so report every time the odometer reached a multiple of 20 kilometres, and press a switch in the steering wheel every time a light on top of the dashboard turned from orange to green or vice versa. The results indicated a marked increase in errors on the first task between the first and second halves of the trip; with the second task the mean, standard deviation and third quartile (Q75%) of reaction times and number of mental blocks increased in the second half of the trip; all these comparisons were statistically significant.

The second study (Näätänen and Summala 1976) concerns the driver's detection of road signs. During a 257 km drive (roughly three hours) drivers were asked to name every traffic sign they saw; these were checked by an observer. The level of detection was extremely high overall, but there were differences between types of

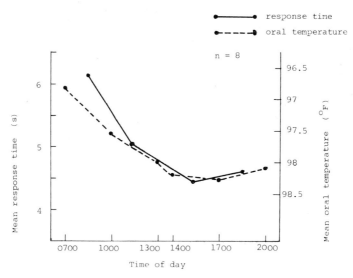

*Figure 4.3.   Response time and oral temperature during a day of prolonged driving.
After Brown et al. (1976).*

sign; as the authors state: 'typically the danger and prohibitory signs were reported with a high frequency (only two unreported signs out of 1431), while most unreported signs were mandatory or informative ones (152 out of 3798). Correspondingly the latter two sign groups and especially the informative signs generally were located at intersections and other difficult places'. There was also a significant change in detection rate as the trip progressed. As Figure 4.4 depicts, following a slight decrease from the first to the second quarter of the trip, there is a progressive increase in the number of unreported signs in the third and fourth quarters; and this is particularly true for inexperienced drivers. A similar finding has been reported by Drory and Shinar (1982), who found that various measures of fatigue were related to the probability of detecting a road-sign on a straight and level road, but not on a hilly winding road.

Hildebrandt *et al.* (1974, 1975) reported the third of these contrasting studies. Train drivers were the subject of this study, but their task is probably sufficiently similar in certain respects to truck driving to argue that certain factors that apply to the former could be important, by analogy, for the latter. The drivers concerned (who worked for the German Federal Railway) have to operate an attention switch whenever they pass a warning signal in order to prevent an automatic compulsive braking system from coming into operation. Thirty seconds before the braking system engages a warning hooter sounds. The results of this study are based on hooter soundings during 6304 working hours of 1000 drivers with an average rate of 3·1 signals per hour; 2238 brakings were also recorded. Both braking and hooter soundings were plotted as a function of time of day, and both frequency curves showed marked peaks at around 0300 hrs and 1500 hrs (the 'post-lunch dip'). The frequency variability of hooter soundings also showed similar peaks indicating that other factors as well as the daily rhythm influenced the increased incidence of errors. One of these is the number of hours worked. Figure 4.5 shows the daily rhythm of the frequency of warning hooter soundings in relation to the number of shift hours

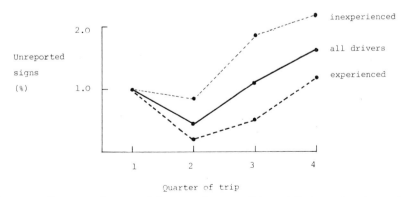

Figure 4.4.   Percentage of unreported signs on each quarter of the route for inexperienced, experienced and all drivers. After Näätänen and Summala (1976).

Fatigue, safety and the truck driver

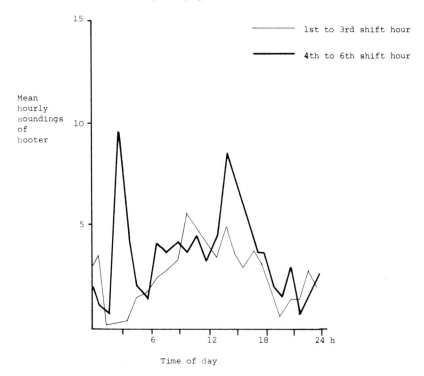

Figure 4.5. Daily course of the mean hourly frequency of sounding of the warning hooter with the first three and second three shift hours. After Hildebrandt et al. *(1974).*

worked at that time. For drivers working the first three hours of the shift there are no early-morning and early-afternoon peaks in error frequency. However, these peaks are strongly pronounced for those who are working their fourth to sixth hour during the critical phases of the circadian rhythm. Another important factor affecting driver errors is the duration of the preceding rest period. Figure 4.6 shows the mean hourly frequency of hooter soundings as a function of duration of previous rest. The number of hooter soundings seems to ascend to a first maximum after 10–16 hours rest; the rate then decreases and the lowest level is reached after 20–24 hours rest; a third maximum is reached during the third day after the previous work shift. After recovery periods of 24 hours or more the daily variation of hooter soundings was found to be quite similar to that generally found for the first three hours of shift. Figure 4.7 shows the relationship between automatic braking and previous rest. Again there is a 'third day' effect, but otherwise the longer the rest the fewer the brakings.

There are several possible reasons that can be advanced to explain why the latter three studies achieved results that appear orderly and consistent with one's expectations about the effects of prolonged performance at a monotonous task, while the bulk of research has achieved essentially negative findings. The light detection tasks of Riemersma *et al.*, Brown *et al.* and Dobbins *et al.* seem in most important

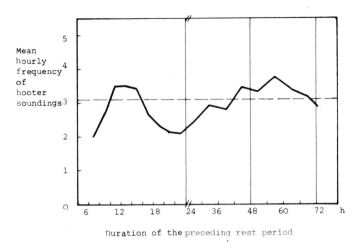

Figure 4.6. Mean hourly frequency of soundings of the hooter warning during the whole shift in relation to the duration of the preceding rest period. After Hildebrandt et al. *(1975).*

respects to be very similar. However, while the first-mentioned was carried out at night, with only one short break, and with presumably very little traffic and few driving incidents, the other two were undertaken during the day with fairly frequent

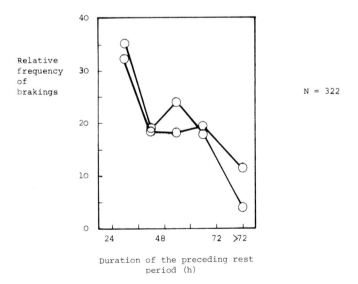

Figure 4.7. Relative frequency of automatic compulsive brakings by locomotive drivers in relation to the duration of the preceding rest period. The two curves represent separate sets of data. After Hildebrandt et al. *(1975).*

rest brakes and performance tests. The importance of circadian factors is suggested by the similar curves for vigilance and body temperature in Brown's study (Figure 4.3). Näätäten and Summala (1976; p. 147) offer several reasons why laboratory simulation studies might more easily exhibit a decline in vigilance than on-the-road studies— there is no danger in leaving the road or having an accident in a laboratory; on the road the driver gets a lot of feedback about his performance; and laboratory tasks tend to be more monotonous. Such arguments are rather undermined by the studies considered here, where performance can be maintained for up to 15 hours on a simulator without noticeable decrement in some functions. The interest provided by the addition of several subsidiary tasks to simulated driving combined with a high degree of subject motivation seem the most appropriate alternative explanations.

Introducing an additional element to a long and tedious task can have the effect of improving performance (and this is indeed the rationale behind introducing warning signals for train drivers). Baker (1960) has demonstrated that increasing the rate of signals by introducing 'false' signals improved the detection rate of the 'true' signals. In similar vein, Lauer and McGonagle (1955) suggest that roadside advertising hoardings may help to keep the driver alert. It is certainly the author's impression from a few field trials with an in-cab vigilance task (detecting a small light on a rear-view mirror) that this task markedly altered the driver's observation patterns, and seemed to maintain his alertness at an unusually high level; indeed he seemed to treat it like a rather novel and curious game. The act of (attempted) measurement can so often totally distort what was intended to be measured.

Hildebrandt *et al.*'s study is the only 'naturalistic' one in that it involved no alterations in the driver's normal working procedure. Näätänen and Summala's study and Riemersma *et al.*'s odometer recording task were somewhat similar in that they emphasized aspects of the driver's task that were already present, rather than introducing extraneous elements. As well as this the signals used in their tasks do not stand out from their background context in the same way as a novel flashing light must do. Possibly therefore these tasks may have been more difficult and required greater concentration than those in the majority of the other studies. Certainly Riemersma *et al.*'s subjects, while initially tending to anticipate the odometer changes during the first driving period, in the second period more often tended to react to the changed signal after it had occurred; also reflecting the difficulty of the whole experimental situation was the opinion of all the subjects that they would have considered it irresponsible to have driven alone and have followed the same procedure as the experiment. Näätänen and Summala's subjects also reportedly found their task exhausting, and indeed the levels of reporting of signs were very much higher than those obtained by Johansson and his colleagues; they used the technique of stopping drivers and asking them to recall the last sign they had passed (Johansson *et al.* 1963, Johansson and Rumar 1966, Johansson and Backlund 1970).

So, although we can generally conclude that under certain circumstances (after prolonged driving, at certain times of the day, etc.) the driver may be less likely to detect some kinds of signals, we can have only a vague notion of the sort of signals that are likely to be missed, and we have virtually no insight into the process of deterioration. We can, however, begin to build up a picture of the driver's attentional system and its susceptibilities to breakdown by considering pathogenic factors other than those normally considered under the 'fatigue' complex, and tasks which, while different from driving, have some features in common with it. The evidence revolves around two crucial aspects of the visual attention process—the importance of peripheral vision and changes in observation patterns.

There is a good evidence to suggest that foveal fixations are at least in part directed by peripheral visual information (Sanders 1966). In a driving context this is well exemplified by Waldram's (1962) simulator study; he found that a pedestrian on the footpath might never be fixated by a driver so long as he continued walking normally; however he attracted an immediate fixation as soon as he simply turned towards the edge of the road. Factors which adversely affect the detection of information in the peripheral visual field include increasing the amount of irrelevant information to the peripheral retina (Mackworth 1965) and, when driving, increasing the level of traffic and task demand (Lee and Triggs 1974). Alcohol may also affect peripheral vision; Buikhuisen and Jongman (1972) found, among other things, that intoxicated subjects missed more traffic incidents at the sides of the screen compared to the centre when watching a film taken from a moving car. Belt's (1969) results can be similarly intepreted (see Rockwell 1972 a); he found in the high alcohol condition (0·08 mg per ml of blood) that there was a much greater concentration of eye fixations near the focus of expansion and fixation duration tended to be greater. That this represents 'perceptual narrowing' or 'tunnel vision' is suggested by the fact that one subject in the high alcohol condition failed to fixate passing cars at all, and that loss of vehicle control seemed to be preceded by sober sampling patterns (indicating perhaps that the perceptual narrowing is adaptive). However, interpretation of eye-movement studies like the latter must proceed with considerable circumspection. Devices for recording eye fixations can indeed record the direction in which the eyes are pointing; however, they cannot indicate what the driver is attending or responding to either in the peripheral or foveal field. Indeed it will be recalled that in the last Section Kaluger and Smith's (1970) results with this technique were interpreted as indicating a degradation in peripheral visual sensitivity and, in contrast to Belt, they found less concentration of fixations in a small area and less viewing time spent near the focus of expansion. There is a great temptation to commit the fallacy of inferring some kind of system degradation from any change in behaviour, ignoring the need for independent validation of the interpretations offered. Research in the area of eye movements while driving is reviewed by Cohen (1978).

Notwithstanding the above strictures, another eye-movement study carries forward the interpretation of the patterning of eye movements. Troy *et al.* (1972) recorded the horizontal saccadic eye movements of helicopter pilots during a cross-country flight. They found that in the later stages of the flight there were fewer single eye movements and an increase in patterned movements (two or more saccades in the same direction). The interpretation offered is as follows: an unusual object detected in the peripheral visual field is shifted to the central (foveal) field; if it is of 'value' to the pilot it is systematically scanned, probably using patterned eye movements; if it has no information value, foveal fixations will shift to other objects in the visual field. Thus a high rate of unit saccades supposedly represents alertness to a large variety of stimuli, and a decline in this possibly reflects decreasing alertness to unexpected stimuli.

As well as being directed by visual information, eye movements are also determined by central processes (Cohen and Hersig 1979); the search and scan processes of a driver or pilot are not entirely determined by momentary changes in stimulation. Wertheim (1978) has constructed a very interesting and stimulating theory around these two (internal and external) determinants of oculomotor activity. He constructed two similar laboratory tracking tasks; in one the tracking stimulus followed a very predictable pattern, in the other the pattern was irregular and unpredictable. Wertheim argues that while in the unpredictable condition the visual sampling pattern would of necessity be determined by moment-to-moment variations in the position of the stimulus (attentive oculomotor control), in the predictable condition the pattern of visual sampling becomes so regular that it begins to follow an internally generated rhythm (intentive oculomotor control). This latter pattern, though matching the tracking stimulus, is not determined by changes in external stimulation. During each tracking task Wertheim recorded his subjects' reaction times to a signal presented either in the centre of the tracking target or in the background of the tracking display. Reaction times to on-target signals were shorter in the predictable than the unpredictable conditions; reaction times to the background signal were the reverse—faster in the unpredictable than the predictable condition. He draws the following analogy between his results and driving: large highways and motorways, where the pattern of visual stimulation is monotonous, will tend to enhance intentive oculomotor control, and the detection of movements outside this pattern (other traffic, road bends) will be impaired. Other roads which are more unpredictable will make for longer overall reaction times due to the enhanced effort inherent in attentive oculomotor control; however, paradoxically, movement changes in the road surroundings will be detected faster. How this theory might clarify the rather contradictory findings of vigilance and driving studies is a matter for further experimentation. The signals in most subsidiary vigilance tasks tend to be in the 'background' in relation to the main tracking task (few have been presented near the focus of expansion, for example). One would thus predict a lengthening in reaction times and a decline in detection rate with the development of an increasingly stable intentive oculomotor control programme. One can only speculate that the driving situations characteristic of the studies by Brown *et al.* (1967) and Dobbins *et al.* (1963) were sufficiently varied and unpredictable to maintain attentive oculomotor control, whereas the opposite was the case in the all-night car driving of Riemersma *et al.* (1976) and the train driving studied by Hildebrandt *et al.* (1974, 1975).

Finally, we come to those studies in which the major emphasis is on speed of response rather than the detection process. Lisper and his colleagues conducted a series of experiments in which drivers were asked to respond (while driving) to a loud (90 dB) tone by either pressing a foot switch or saying 'bom' (Lisper and Eriksson 1980, Lisper *et al.* 1971, 1973, 1979, Laurell and Lisper 1976). Typically the average inter-signal interval was 50 seconds, ranging between 10 seconds and 2 minutes. Such an effective 'foreperiod' is much longer than in typical classical reaction time studies (Woodworth and Schlosberg 1954), thus possibly involving changes in 'readiness' or 'attention'. But because of the more direct 'coupling' between an auditory stimulus and the brain (compared to the visual system which is directional), and the fairly high frequency of signals, this set of experiments has much more in common with the classic reaction time procedure than the 'vigilance' studies discussed earlier. The 1971, 1973 and 1976 studies of Lisper and colleagues comprised four, three and two hours of driving respectively. Increases in reaction time (RT), were found in all three studies as driving time progressed. In the 1971 study no differences were found between RT while driving during the day, during the night, or during a day following a night without sleep. In the 1979 study circadian variation was found to have only a minor influence on RT, while in the 1980 study having a meal seemed to prevent a deterioration of RT with time driving (eight hours in total), though having a rest break of one hour after four hours did not. In the 1973 study, only inexperienced drivers showed an increase in RT; experienced drivers showed a slight decrease (the difference in trend between these was the only significant effect). The 1976 study compared RT while driving with RT while being a passenger and with RT while sitting in a stationary car. Only in the driving condition was there a significant increase over time, indicating that this was due to driving rather than the performance of the RT rest itself. The increases in RT were around 100 milliseconds in the 1971 study but much less in the other studies. Dureman and Boden (1972), using a very similar procedure, though in a simulator rather than in a car, report no significant RT changes over a 4-hour session.

Various studies have recorded reaction times not during driving, but using a before-and-after type design. Lahy (1937) followed a truck on a 3-day journey with a mobile 'camion laboratoire' testing his subjects at various intervals with various psychophysical and physiological tests, including simple reaction time. His descriptive analysis of results shows a clear relationship between RT and driving (with, of course, longer RT after driving). Jones *et al.* (1941) administered a large battery of tests to nearly 900 drivers in a number of cities in the USA; scores on the tests were related to the number of hours driven before the testing; two of the tests which tended to be performed worse by those who had driven for longer periods were simple reaction time and reaction co-ordination time (the time required to move a pencil from one hole to another). Johnson (1945) severely criticizes this study both for the inadequacy of its design and the irrelevance of its tests to driving. Yajima *et al.* (1976) measured simple and complex reaction times of drivers every 1–2 hours during a series of 7–10 hour drives. Results are presented in a very vague and inadequate fashion, and suggest very little consistency between subjects, but where a change in RT occurred as the trips progressed it tended to be a lengthening of RT.

What validity have these measures of reaction time in terms of safe or skilful driving? Lisper (1975) reports a study in which performance on the subsidiary task was compared to parallel measurement of detection distances to objects on the roadside. The correlation between changes in these variables as a function of time was −0·75. Concerning laboratory reaction time measures, Babarik (1968) found that taxi drivers with faster reaction times tended also to have lower overall accident rates. It is undoubtedly true that there will be some circumstances in which a slowing of reactions by one-tenth of a second may have important consequences for accident avoidance or severity reduction, and thus these reaction time studies have considerable face validity. However, and notwithstanding Lisper's and Barbarik's results, it seems very important to evaluate reaction time in terms of the driver's total response to an emergency situation: his detection and evaluation of a potential accident situation, his selection of an appropriate response, and his speed of execution of that response. Small changes in response latency in a simple reaction time situation may be entirely trivial when considered in terms of the speed and adequacy of the overall response pattern. And indeed this is borne out by Currie (1969). Using model cars and roadway to mock up potential accident situations—overtaking with an oncoming car, lane violations, cross-roads with ambiguous rights of way—he found that 'safe' drivers (with low accident frequency) tended to perceive the danger in some of these situations faster than drivers with a greater history of accidents; however, there was no relation between simple reaction time and accident experience, nor between simple reaction time and detection times for the potential accident situations.

Thus, what can be concluded about the driver's detection of and speed of response to events, situations, signals, signs or objects, whether they represent a danger of accident or penalty, or convey straightforward road or traffic information? While by far the majority of both simulator and on-the-road studies have failed to find that the driver's detection of a discrete signal is any less likely, or slower, after long periods of driving, there are a few studies which suggest circumstances where this finding may not hold. These latter studies have each exhibited one or more of the following features which may have contributed to the decline in performance accuracy, and which were not in general characteristic of the former (no decrement) studies: driving at night; driving for prolonged periods without rest; driving along a monotonous route; performing a vigilance task that requires a high level of concentration and effort; or one which is part of the normal activity of the driver rather than being a novel 'artificial' task. One study of train drivers has been particularly illuminating in identifying the interacting effects of hours on shift, time of day, and the duration of previous rest on error rate. The theory of signal detection does not seem to be an appropriate model in accounting for these results; rather, an explanation might lie in changes in the patterning of oculomotor activity. It is suggested that, particularly along monotonous routes, oculomotor activity may become increasingly determined by an internal (intentive) control pattern, and the driver may become less responsive to events in the visual background. This is possibly associated with a fall in peripheral visual sensitivity, though there is no direct evidence for this. There is evidence that a driver's reaction time is likely to be slower after a number of hours of driving.

Although there has been some attempt to validate reaction time measures against accident rates and detection distances, the relationship between either the reaction time of the vigilance tests in this section, and the driver's detection, evaluation and response to crucial traffic situations, or other events or circumstances related to his safety, remains a matter for speculation.

4.4. *Interacting with other vehicles*

The account so far of the driver's task would be fine if he or she were in the only vehicle on the road. This section explores the skills involved in interacting with other vehicles in such manoeuvres as following, overtaking and passing, and merging into a stream of traffic from another lane or slip road. What little evidence there is that these skills may be susceptible to degradation under 'fatiguing' conditions will be evaluated, and we will discuss the merits of construing these changes in performance in terms of risk-taking, rather than as a direct function of perceptual/motor processes.

One of the most frequent multi-vehicle interactions occurs when the lane ahead is obstructed by a slower-moving vehicle, requiring the following driver to adjust his speed, or overtake. Given an otherwise clear road and a sufficient speed differential the easiest manoeuvre is a 'flying pass'. Forbes and Matson (1939) found that drivers they observed would not usually overtake unless the initial speed differences was at least six or seven m.p.h. However, drivers rarely follow other vehicles for any length of time except when constrained by dense traffic or the narrowness or twistiness of the road. Rockwell and Snider (1965) found this to be particularly true of professional truck drivers who served as experimental subjects in a naturalistic study of following performance; incidents of following were rare and transient, with the drivers tending to minimize the influence of other vehicles on their own speed control by either overtaking or falling well back behind other traffic. If the driver does choose to follow another vehicle, the way he does so will obviously depend upon whether, for example, he is actively seeking an opportunity to pass, or content to maintain his position in a long stream of traffic.

When following, the driver has continuously to monitor the lead vehicle's (and perhaps, in dense traffic, several vehicles' ahead) distance and relative velocity to maintain a safe gap. Similar judgements are required in other gap-judging situations, such as overtaking in the face of an oncoming car, changing lanes on a busy multi-lane highway, or joining one from a slip-road, and turning across a stream of traffic. The sort of information that is required is similar in all these circumstances, though the boundaries of what constitute safe gaps are obviously determined by differences in the direction of travel, relative speeds, etc. appropriate to the situation. How the driver performs is a function both of his accuracy in sensing the important elements in the situation, and of the confidence he has in making his judgement of the size of the gap and the margin of error he is prepared to accept in carrying out the manoeuvre.

Consider first of all the following situation: three phases can be identified—the initial approach behind the lead vehicle, turning into what Michaels (1965) calls 'steady-state' following (where it is important to detect and respond to small changes in the relative position and movement of the two vehicles), and finally, whatever manoeuvre terminates the following. Initially, of course, it will not be possible to sense directly the speed of the lead vehicle. As far as the following driver is concerned, the gap is bounded by the distance to it, and he will infer the relative velocity of his and the lead vehicle from successive appreciations of the distance between them. Important cues to distance will of course include the comparison of the apparent size of the vehicle to its known size, as well as cues related to the geometry of the road and its surrounds (gradients of linear perspective and texture, relative upward location in the visual field and so on). At closer distances, the relative velocity of the lead vehicle can be sensed directly, the primary cue being the rate of change of visual angle subtended by the lead vehicle. Figure 4.8 depicts the general relationship between distance and the threshold of relative velocity detection based upon psychophysical studies both in the laboratory and the field (see Michaels 1965). There are, of course, great individual differences in threshold—and this is a prominent feature of Janssen *et al.*'s (1976) results. There are, also, other cues to relative velocity; for example, at night, the apparent size and brightness of tail-lights, though Janssen *et al.* have found these cues to be less important than changing visual angle. The pitch of the lead vehicle as it accelerates and decelerates may also provide an indication of a change in relative velocity.

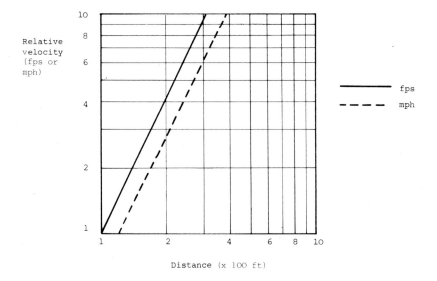

Figure 4.8.   The minimum relative velocity which is detectable to a following driver. After Michaels (1965).

Michaels (1965) gives a thorough theoretical account of perceptual factors in car-following, and he summarizes his model as follows:

... there are essentially three distinct processes that the driver must use to ultimately achieve stable following. In the initial phase of overtaking where angular velocity is below threshold, the driver employs a detection of change in distance to determine the existence of overtaking. This process is a simple binary judgement and hence overtaking is independent of relative velocity. The second process involves detection by the following driver of the angular velocity of the lead vehicle which is a function of both relative velocity and distance. The driver then responds to this rate of change by reducing his speed sufficiently to keep the angular velocity at or near his absolute threshold of detection.... Finally, since the driver requires a certain minimum headway in order to insure vehicle steering and speed control he must finally adjust his relative velocity to zero at the appropriate headway. Using this headway as a reference the driver will, because of the relative velocity relationships, tend to overshoot and then recover.

Steady-state following, Michaels defines in his paper, is when the angular velocity of the lead vehicle is below the threshold of detection of the following driver, and the latter is thus adjusting his following speed and distance on the basis of just-noticeable differences in distance separation between the two vehicles.

This model does generally seem to fit the way drivers actually behave in following situations. Perchonok and Seguin (1964), in a naturalistic study of traffic on a multi-lane urban expressway, found that relative velocity does have a small but significant influence on the acceleration of following drivers, and this influence declines as distance headways increase. However, their data suggest that relative velocity sensing is possibly not so important as is implied in Michaels's model—its maximum influence is in headways of around 70 feet; much below that (less than about 50 feet) drivers become decreasingly responsive to relative speed and increasingly responsive to vehicle spacing; over 100 feet relative speed exerts little influence on the following driver. Rockwell and Snider (1965) investigating a controlled leader-follower situation found that following drivers resist duplicating small changes in lead-vehicle velocity unless forced by short time headways. Their drivers showed reluctance generally to change their velocity, delaying changes as long as possible, preferring lead car velocity variance to be absorbed in headway variance. In view of the difficulty of the perceptual discriminations involved in monitoring relative velocity and small changes in headway, it is not surprising that drivers attempt to maintain their own pace as much as possible and to avoid following whenever traffic permits. Indeed, Perchonok and Sequin found that the three stimulus variables they studied (distance headway, relative speed and the ratio of the two) had very little overall influence on drivers' behaviour, accounting for only about 15% of its variation.

How accurate are drivers in judging headways and headway change? Rockwell (1972 b) summarizes some of the evidence: absolute headway estimation is very poor in most driving situations, with errors ranging from 20% to 100%. Data on actual headway production are similar—subjects instructed to maintain headways of 200 feet behind a lead car travelling at 60 m.p.h. with small variations follow at distances ranging between 50 and 300 feet.

However, this considerable inaccuracy of distance estimation is balanced by quite a high degree of sensitivity to the sign of relative motion between the two vehicles; furthermore, subjects show a consistent bias in favour of judging negative relative motion, which is of course in the direction of increased safety (Olson *et al.* 1961, Evans and Rothery 1974). This bias is quite explicable in terms of the optical geometry of the situation. As Michaels (1965) points out, when the lead vehicle decelerates, the visual angle it subtends will increase exponentially due both to the deceleration itself and the decreasing distance; when the lead vehicle accelerates the angular velocity (rate of change of visual angle) initially increases, but as separation also increases it will decrease exponentially and drop below threshold. These relations are shown in Figure 4.9. Given a high degree of accuracy in detecting the lead vehicle decelerations, the maintenance of safe following distances is largely a function of driver reaction time and, more importantly, time lost because of inattention (Lehman and Fox 1967). These authors also point out that safe driving distance is not so much dependent on absolute values of speed and deceleration as it is on speed changes and relative decelerations of the two vehicles. There is, however, a considerable possibility of potentially dangerous error in the driver's ability to judge not the sign of relative motion but its magnitude, and in this context Rockwell and Snider (1965) found that lead vehicle acceleration rate, particularly negative acceleration, had little or no effect on headway change detection.

Drivers will of course adopt different following strategies under different circumstances. Obviously they will follow closely when preparing to overtake; Näätänen and Summala (1976; p. 216) report that a prohibition sign forbidding overtaking reduces the number of very short headways. Sumner and Baguley (1978) report that close following by heavy goods vehicles is common on motorways, and that there is more close following as speeds increase. This probably represents a

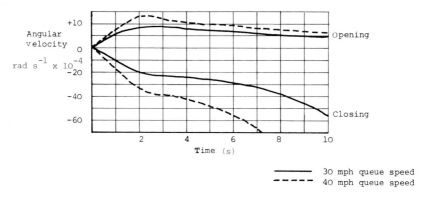

Figure 4.9. The change in angular velocity for a following driver when a lead vehicle accelerates at $3\,ft\,s^{-2}$ for $2\,s$, and then maintains a relative velocity of $-6\,ft\,s^{-1}$. After Michaels (1965).

compensation by the drivers for their vehicle's poor acceleration and serves to maintain their ability to overtake. Wright and Sleight (1962) found that drivers followed more closely under an 'emergency' than a 'safety' instructional set, and more closely at 30 m.p.h. than 50 m.p.h. There is also a potential speed adaptation effect; Denton (1967) has reported investigations of the phenomenon that after a period of high-speed driving a reduced speed seems much slower than it really is. Irving (1973) reports that this speed misjudgement can lead to the adoption of shorter following distances than those when the adaptation to the lower speed is complete.

Until recently there had been no investigations of the way in which a driver's following performance or strategy might change over a prolonged period of driving, or in other circumstances associated with fatigue. Rockwell and Snider (1965) found a great consistency in the performance of professional drivers during four hours of driving with intermittent periods when the driver was instructed to follow another vehicle. Lerner *et al.* (1964) obtained a similar result after two hours of continuous following. However, one small clue that under certain circumstances drivers can find themselves in a very dangerous situation was provided by Edmondson and Oldman (1974), who reported various anecdotal accounts by shiftworking truck drivers of how occasionally a driver, presumably very drowsy, might follow very closely behind another truck—until warned off by a sharp jab on the brakes! It is hard to provide a good explanation for this phenomenon; perhaps at night, on a long monotonous route, the tail lights of another truck may provide a tracking stimulus requiring less visual effort than the road ahead to a man who is very drowsy, has difficulty in keeping his eyes open, who may be in a semi-sleep or hypnagogic state, where his perceptions of speed, distance, and the riskiness of his situation may be severely impaired.

Since then an extensive experimental investigation of following episodes in normal traffic over prolonged driving periods occurring at different times of the day has been conducted by Fuller and McDonald (Fuller 1978, 1980, 1981, McDonald 1978). In this study each subject was required to drive for 11 hours per day for four consecutive days; the driving sessions started at either 0900 or 1500 hrs. The results did not support a hypothesis of progressive impairment of performance with time. Significant time-related changes in the average headway (measured by time, not distance) and the variability in headway adopted by drivers were found, but these were most plausibly explained in terms of the driver adopting different following strategies under different road and traffic conditions. Thus in urban areas where traffic tends to be denser and there is a greater probability of the lead vehicle decelerating or stopping at any particular time, the driver tends to maintain a longer time headway between him and the vehicle in front, whereas in rural areas, where traffic tends to be faster, sparser and travelling at a more predictable speed, shorter headways are found. A rather different pattern of results was obtained when only short time headways (of less than two seconds) were analysed separately; once again the evidence did not support a fatigue hypothesis and the most likely explanation again probably lies in the pattern of traffic interactions occurring throughout the experimental sessions, though no adequate analysis of this was possible. There were some differences between groups in that younger drivers on the late shift tended to follow closely more often than the other drivers; they may have been overtaking and passing more often, or following other vehicles in order to counteract the boredom which they tended to feel more than the other groups.

A development of this study involved convoy driving (continuous following of another vehicle) under the same operational conditons. Under these circumstances drivers more often showed symptoms of fatigue than in the earlier study—for example, they more frequently expressed a preference for stopping earlier. Older drivers on the late shift were affected more than other shift and driver combinations. However, this pattern of fatigue was accompanied by longer rather than shorter headways, which Fuller interprets as adaptation by the drivers to their deteriorating subjective state—longer safer headways are adopted to compensate for fatigue. In a further study, manipulating the predictability of the time the driving shift started and its duration had no effect on performance or subjective fatigue (Fuller 1983).

Two things are clear from these studies: firstly, that the interpretation of a measure of performance like time headway needs to be sensitive to a complex pattern of road and traffic interactions; and secondly, one should not necessarily expect a simple effect of hours of driving on performance. In the first experiment none of the drivers expressed any great difficulty in fulfilling any of its requirements; what they were asked to do differed little from their normal work (and may have been easier in some respects). Thus it is not surprising that nothing like a deterioration in performance was found. In the later experiments such relatively minor symptoms of fatigue which appeared seemed to be well compensated for by increasing safety margins.

In a manner similar to the car-following situation, human sensory capabilities are not particularly well suited to the task of overtaking and passing other vehicles. It is likely that at most distances at which the decision whether or not to overtake has to be made, the relative speed of an oncoming vehicle will be close to or below threshold. Both Rockwell and Snider (1965) and Bjorkman (1963) found that drivers were to a limited extent sensitive to the speed of an oncoming car in fairly typical conditions, but it is clear that the major cue that the overtaking driver has of the size of the overtaking gap is his estimate of the distance to the oncoming car. This is reflected in Bjorkman's finding that subjects' estimates of the meeting point between their own and an oncoming vehicle tended to be biased towards the midpoint between the two vehicles—the subjects having presumably made the assumption that both vehicles are travelling at the same speed. Bjorkman did however find that feedback improved accuracy—suggesting that, with training, drivers could make better use of what little relative velocity information is available to them.

Some overtaking gaps are defined by terms of distance—to a bend or narrowing of the road, or road markings, for example—rather than by the speed and distance of an oncoming car (which is most easily expressed as a time gap). The ability of drivers to judge either kind of gap is very limited. Thus Gordon and Mast (1968), comparing estimates of how much distance was required to execute a pass with the distances actually taken, found, for example, that 60% of estimates of the distance required to pass a 50 m.p.h. car were too small. Generally errors ranged between 30 and 50% of the actual distance required, and the amount of underestimation increased as the speed increased—as also does the actual distance required to pass; the rate of error was an accelerating function of speed. Knowing the performance capabilities of one's own car is important, and subjects driving an unfamiliar car had a much higher error rate. In a similar experiment, Jones and Heimstra (1964) compared subjects estimates of minimum time gaps (to an oncoming car) with their actual performance: 30% of estimates were too small.

This difficulty in estimating the size of overtaking gaps is perhaps not surprising considering the complex relationship between the elements involved. Drivers seem to be reasonably accurate in assessing their own speed (Norling 1963), and given a reasonable familiarity with their own car can assess its ability to outpace the lead car. However, the difficulty arises in having to compute this potential relative velocity, or relative acceleration between their own and the lead car, in terms of either the absolute distance covered along the roadway by both cars, or, with an oncoming car, the speed and distance relationships between the subject's own car and it. Figure 4.10 is a graphical representation of the time/distance relationship involved in overtaking in the face of an oncoming vehicle.

As would be expected, the inaccuracy of drivers' powers of estimation is matched by their variability in performance in both semi-naturalistic studies in normal traffic and more controlled experiments on test tracks (Matson and Forbes 1938, Forbes and Matson 1939, Crawford 1963, Quenault 1973, Rumar and Berggrund 1973).

Rumar and Berggrund's study is a very good example of a controlled investigation on test tracks. They asked their subjects not to overtake until the last safe moment. Reflecting the considerable difficulty in judging this minimum safe gap is the following pattern of results: far from beginning their overtaking manoeuvre at a greater distance from the oncoming car when it was approaching faster, subjects tended to, if anything, do the opposite. There was a huge range of 'safe distances' to the oncoming car that the overtaking drivers selected—a range of up to 1000 metres. Difficulty in estimating oncoming car speeds was also reflected in the great variation in time margin left at the end of the manoeuvre between the three oncoming car speeds (very much less with faster speeds). Time margins left between the two speed conditions selected for the overtaken car (and thus the overtaking car before the manoeuvre) showed no great difference, indicating a good appreciation by the drivers of the acceleration capacity of their own car. Thus while most drivers in this experiment were able to safely accomplish the manoeuvre with the lead car travelling at 80 k.p.h. and the oncoming car at 60 k.p.h. only about half managed to do so with respective speeds of 80 and 100 k.p.h., those drivers having problems in the former situation having even larger problems in the latter. On the other hand it was found that drivers were fairly good at estimating the safety margin they had left—thus at least they get some feedback on their performance, though maybe not early enough to effectively correct many mistakes.

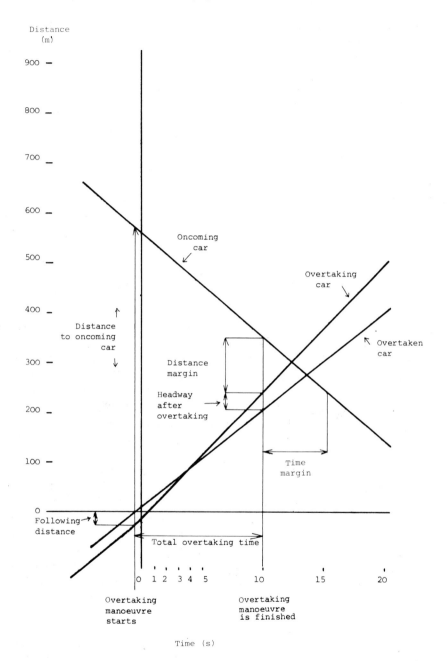

Figure 4.10.   *Time vs. distance diagram of an overtaking manoeuvre.*
After Rumar and Berggrund (1973).

Of course this experimental set-up is not typical of all overtaking situations. Drivers usually prefer to undertake 'flying passes' rather than start from the same speed as the lead car. Also, in the normal situation the driver does have the opportunity to change his mind, and many accidents are averted by the other cars in the situation taking avoiding action (though this is not always appropriate). However, this general pattern of findings is repeated in studies in normal traffic. Forbes and Matson (1939) found that 10% of overtakings involved a negative clearance time, while half of the rejected gaps were sufficient for passing. Quenault (1973) confirms the pattern, finding a large proportion of overtakings being rated unsafe due to the oncoming car being too close. Rumar and Berggrund's subjects comprised also part of Quenault's sample in his observational study of drivers' performance. Unfortunately none of the ratings of safety of driving or of overtaking seemed to be related to the drivers' safety margins in the experimental situations. Quenault's analysis is directed towards identifying stable personality differences between drivers. Thus the observed patterns of skill tend to be interpreted in this light—as reflecting underlying personality differences—and little or no emphasis is put on the range of situational variables that could affect the driver's behaviour. Because of this slant in analysis of his results, their usefulness is somewhat minimized.

Another factor that adds to the inherent instability of the overtaking situation is the finding that the time it takes the driver to decide whether or not to overtake increases with the greater urgency of the decision, i.e., when the time or distance gap available to the driver is less. Crawford (1963) calculated a threshold range for gap acceptance between the smallest accepted and largest rejected intervals; all cases when a subject changed his mind and terminated an overtaking manoeuvre occurred within this range, and generally as the gap size decreased the time taken to respond increased (though large response times were also characteristic of large gaps where there was no time pressure on the subject). De Kock (1968), in a simulated overtaking situation, confirms the evidence that a more difficult decision results in a longer decision time.

It is very fortunate that we can compare this theoretical and experimental analysis of the difficulties and dangers of the overtaking manoeuvre with data on the manoeuvres of vehicles involved in accidents. Lewrenz and Pittrich (1973) present a very comprehensive analysis of circumstances surrounding 4600 overtaking accidents in five European countries. This confirms that endangering oncoming traffic (probably largely due to perceptual problems) is one of the major causes of overtaking accidents. At least as important (more so in terms of frequency and severity of injury and number of deaths) is the behaviour of the overtaken driver in causing accidents—particularly careless changes of direction, reflecting, most likely, inadequate observation of the total road space. (It is interesting in this regard to note that Wilde and Neipold (1973) report that drivers are fairly realistic in estimating the danger in overtaking situations, and that their estimates of danger are not so much related to the distance available for passing, as to uncertainty over how the vehicle being passed will behave.) Other results of Lewrenz and Pittrich's analysis indicate the need for communication between drivers to eliminate accidents due to the overtaking car cutting in on the overtaken one, or the oncoming car failing to stop in time, both situations arising, one would suppose, very often as a result of initial misjudgements by the overtaking driver of the gap available.

Other gap-acceptance situations that involve many of the same perceptual problems as overtaking are turning across a lane of moving traffic, and merging into a stream of moving traffic from either a stationary position or from moving down a separate driving lane. In both of these situations the driver has to assess relative rates of motion of other vehicles; however, neither involves a critical gap size nearly as large as that involved in overtaking and passing. Thus in most intersecting and merging situations it is possible to directly sense the speed of the other vehicles which bound the gap, and, as Gibbs (1968) suggests, visual cues to speed will include the yaw and roll of the other vehicle and its general response to road surface irregularities, as well as the rate of change in visual angle subtended by it. Perception of speed is not very accurate, however, and while Gibbs (1968) suggests that at intersections drivers simply estimate gaps in terms of time before the oncoming vehicle arrives, data collected by Bottom and Ashworth (1973) imply a more complex process: "either the drivers are basing their decisions on fixed distance and modifying this with regards to the speed of the oncoming vehicle or . . . the driver is basing his decisions essentially on time gaps but is repeatedly under-estimating the speeds of fast vehicles and/or overestimating the speeds of slower vehicles." In lane-merging, where both the gap and the gap-accepting vehicles are moving, it is likely that the driver is concerned about the ability of the following driver to accommodate him by decelerating before his headway is exhausted, and again does not simply rely on a simple time gap estimation (Hurst *et al.* 1968).

Particularly when one considers the rather low level of accuracy with which drivers can judge the gaps involved in overtaking, merging, and at intersections, and the distance to a lead vehicle in following, it is apparent that there must be considerable uncertainty over the most appropriate action to take—to accept or reject the gap, or to adjust headway. The action that the driver does adopt will depend on how certain he is that his judgement conforms to the realities of the situation, and the level of certainty upon which he is prepared to act—the risk he is prepared to take. Hurst (1964), Edwards (1968), Näätänen and Summala (1974, 1976) and Wilde (1982) discuss models of the driver as risk-taker. In terms of accidents, as Edwards points out, the driver is dealing with a very small probability of a very great penalty; Michon (1973), however, contends that the driver is not capable of appreciating, in any meaningful sense, probabilities of that magnitude. Although the subjective estimation of risk is probably a very simple unsophisticated procedure, it seems likely that some conception of risk taking can be usefully applied to most driving situations. Senders *et al.* (1966, 1967), for example, have discussed the drivers' temporal information seeking patterns in these terms. However, the driving situation which has attracted most experimental attention in this context has been driving through a fixed gap on the roadway. This is obviously a much simpler situation than accepting or rejecting an overtaking gap, but it is not necessarily of trivial importance; Lewrenz and Pittrich

(1973) point out that the dangers inherent in a very similar type of judgement: overtaking with insufficient lateral distance from the overtaken vehicle is a causative factor in a high proportion of overtaking accidents.

Cohen and his colleagues asked bus drivers to estimate how many times out of five they could negotiate their bus through various gap sizes, comparing experienced with inexperienced drivers, and the effects of alcohol (Cohen *et al.* 1956, 1958). Results are interpreted in terms of changes in assessment of the subjective probability of passing the gap; thus Cohen (1962) suggests that, both slightly drunk and sober, a driver will attempt gaps of the same subjective probability—but the alcohol will influence him to assess a narrower gap as being passable with the same probability as a wider gap when sober. Zwahlen (1973 a, b) suggests that a similar task can provide a fairly stable estimate of a subject's risk-taking propensity. The generalizability of these findings to other driving situations has yet to be proven, however. Judging a static gap in an experimental situation, where the consequences of failure are, at worst, some social embarrassment, is not the same problem as that faced by a driver under time pressure faced with a marginal overtaking gap with a 40-ton truck at the other end of it.

Considering now prolonged driving, shiftworking, deprivation of sleep, or other circumstances associated with 'fatigue', it is apparent that any change for the worse in the driver's decisions concerning following, overtaking, lane merging, etc., might be imputed to either a perceptual change making him less sensitive to crucial information, or to a motivational change, making him prepared to accept greater objective risks (whether or not his subjective monitoring of risk agrees with this or not). Few studies however have studied these types of driving decisions over prolonged periods. Brown *et al.* (1970) investigated the riskiness of overtaking manoeuvres during four consecutive three-hour continuous driving sessions with short breaks between. Riskiness was judged according to two criteria: whether or not forward visibility was restricted and whether or not the manoeuvre was possible only if another vehicle changed speed. Results showed an increase in the proportion of risky to safe overtakings from the first to the fourth session. Brown and his colleagues suggest that this increase reflects a change in the subject's decision criterion in favour of greater risk, rather than being due to perceptual factors, arguing on the basis that both performance accuracy on a subsidiary visual vigilance task and 'arousal level' (oral temperature) increased during this period.

There is some evidence to support the view that fatiguing conditions can lead subjects to choose riskier courses of action. Holding and his colleagues (Shingledecker and Holding 1974, Barth *et al.*, undated report) investigated the effects of risk-taking tests of both continuous work for 24–32 hours on a demanding battery of mental and perceptual tasks, and of hard physical work on a treadmill or bicycle ergometer. The risk-taking tests involved the choice of several alternative routes to a single goal 'with each route requiring a different amount of perceptual or motor effort and each route having a different probability of leading to success' (Barth *et al.*, *op. cit.*). After the fatiguing tasks (and the subjects did rate themselves as being 'fatigued'), choices tended to be towards alternatives requiring less effort despite the lower probability of success: a riskier strategy (physical fatigue did not, incidentally, lead to a riskier strategy on a perceptual risk taking test).

A motivational account of any changes associated with fatigue that may occur in drivers' decisions in a variety of traffic situations is a lot more compelling than an account based upon perceptual changes—though in practice it is extremely difficult (and perhaps not very useful) to draw sharp distinctions between perceptual, judgemental and decisional components of the total action. The driver does have considerable freedom in the pacing of his work (though the commercial driver will often be working within fairly tight constraints of schedules); thus, while he is working within the limits of his sensory abilities, what he actually does is determined by his confidence in the accuracy of his judgement, and the level of risk he is prepared to take in executing the manoeuvre. Certainly, from a commonsense point of view, considering the experience of fatigue, and perhaps also drowsiness, to be characterized by an aversion towards continued activity, and a desire to be out of the unpleasant situation that is causing discomfort or preventing sleep, the 'subjective utility' to be gained by accepting a certain overtaking gap, for example, will be increased when the driver is tired and construes accepting the gap as helping to shorten the driving time. On the other hand gap acceptance decisions are difficult and involve considerable uncertainty. 'A feeling of unwillingness and inadequacy for activity', part of Bartley and Chute's (1947; p. 54) definition of fatigue, might just as easily show itself in avoiding, wherever possible, traffic decisions, or if they are unavoidable, reducing the effort involved by accepting only easy ones, or by being less concerned about the accuracy of the perceptual judgements involved, or of assessing the risks.

It is unfortunately not possible to fill out this commonsense level of analysis with more comprehensive empirical and theoretical support—the basis for such an analysis does not exist. However, some small confirmation of some of these general ideas can be had from a few studies that have used observers to assess fairly general aspects of driving skill. Brown (1976 b) had experienced police drivers assess the performance of six non-professional drivers over a short course in city traffic, both before and after seven hours of virtually continuous driving. 'Perceptual skills' (anticipation of traffic changes, positioning on the road, speed in reacting to unexpected demands) and 'courtesy shown towards other road users' (use of turning signals, conforming to speed restrictions, precedence given to other traffic at junctions) were together rated worse after continued driving, while motor skills (competence in using hand and foot controls, acceleration imposed on the vehicle) were not. Both McFarland and Moseley (1954) and Harris *et al.* (1972) had observers rate driving errors (ranging from errors concerning speeding, tailgating and signalling to various types of error of judgement) during normal commercial driving runs, the former study over one 240-mile bus run, the latter throughout a range of different truck and bus operations. Although in some cases there seemed to be evidence of an increasing trend in errors after prolonged periods of driving, and of an improvement in performance following rest breaks, these trends were just as often contradicted. Unfortunately, in both of these studies the data are presented as simple frequency counts of overall errors as a function of time, and so no more detailed qualitative analysis is possible.

In conclusion, therefore, the evidence testifies to the inadequacy of the driver's sensory capabilities in enabling him to sense accurately the movement of other vehicles in relation to himself. Thus there is a built-in instability in any multi-vehicle interaction. The major problem is the sensing of relative velocity, the chief cue to which in most situations is the rate of change of visual angle subtended by the other vehicle; this is not a very accurate cue and in many interactions will be near or below the driver's perception threshold. Thus, in following other vehicles, relative velocity is detectable only at intermediate distances (or with large speed differences) and, while drivers are accurate in detecting the sign of relative motion (particularly negative relative velocity), they are not good at detecting its rate; they tend to avoid duplicating changes in speed of the lead vehicle as much as possible, and in fact avoid following other vehicles whenever traffic and the road permit. Drivers are only minimally sensitive to the speed of an oncoming car when overtaking and passing; they are very inaccurate in judging both distance and time gaps required for overtaking, and both controlled studies of overtaking in normal traffic and test tracks, and accident statistics point to the great variability and inadequacy of the driver's performance. Although not so extensively studied, other gap-acceptance situations share essentially the same problems. What little evidence there is suggests that prolonged driving may be associated with a greater tendency to take risks when overtaking, and a general deterioration of skills involved in interacting with other road-users. However, despite some anecdotal evidence which suggests that drowsy drivers may occasionally follow dangerously close behind other vehicles, one should not necessarily expect this to occur as a result of a relatively normal long day of driving; on the contrary, drivers may tend to compensate for fatigue by adopting safer margins. It seems more appropriate to explain the evidence for deterioration in complex driving skills in terms of the driver's appraisal of the risks involved (either assessing the perceptual evidence differently, or accepting a different level of 'subjective risk') rather than in terms of changes in perceptual sensitivity.

4.5. *The mental demands of driving*

The research quoted in the previous sections of this chapter has amply demonstrated that most drivers, under most circumstances, are well able to continue driving for prolonged periods without their performance deteriorating in any noticeable way. The task of driving does not normally put an excessive demand on the driver's perceptual, attentional cognitive or motor abilities. However, there are emergency occasions which do tax his skill to the limit. According to Brown (1962 a) a good driver is "one who maintains sufficient spare capacity to deal with an unexpected but possible event"; this 'spare capacity' is presumably maintained by altering the speed of driving, accepting a certain margin of error in traffic decisions, etc. Thus, it is argued, during normal driving, direct measures of performance will not be sensitive to fluctuations in the demands of the driving task or to whatever sort of impairment is involved in driving 'fatigue'—such changes will be absorbed in the driver's 'spare mental capacity'. However, it is possible, according to this theory, to measure changes in 'spare capacity' associated with task load or fatigue by means of a subsidiary task which the subject is asked to do concurrently with the main task (driving). The subject is instructed to perform the primary task as well as possible at all times and do the secondary task when he can; thus, hopefully, the primary task will be allocated a greater or lesser amount of 'mental capacity' according to the demands of the main task (Brown 1964).

In Brown and Poulton's (1961) words: "the theoretical basis of the technique is ... the assumption that there is a limit to the rate at which an operator can deal with information—in other words he has a limited 'channel capacity'. When the demands of the primary and subsidiary tasks together exceed this limit errors must occur". In more concentrated jargonese: "the human operator is fundamentally a one-channel data-processing system having limited capacity" (Brown 1964). Channel capacity is not fixed but is related to, for example, the general level of arousal of the individual— thus the result of an increase in task demand may be in some circumstances an increase in arousal (increasing total channel capacity) rather than a decrease in 'spare capacity'. Macdonald and Cameron (1973) therefore suggest that, in some individuals, measures of arousal may provide a more sensitive measure of primary task difficulty than performance on the secondary task; in others the secondary task may be the more sensitive measure. The question of arousal and task load will be further discussed in the following chapter.

There are various requirements that, theoretically, a subsidiary task should fulfil in order to be an adequate measure of reserve capacity. Obviously it should be easily discriminable from the main task and preferably should use sensory and motor modalities other than those used in the primary task, thus ensuring that changes in performance reflect central processing rather than perceptual or motor overload. Stimulus–response compatibility should be high within the secondary task but low between it and the primary task. An unpaced task is less likely to interfere with the primary task, but is more susceptible to practice and motivational fluctuations; a paced task should exceed the subject's capacity and thus ensure errors. Performance units should be brief, easily 'shed', and responsive to brief fluctuations in task difficulty. The task should appear less 'important' than the primary task and should not 'challenge' the subjects in the way that a mental arithmetic test may do. Curry *et al.* (1975) and Macdonald and Cameron (1973) discuss these and other considerations in greater detail.

There are some problems with the secondary task method. Because reserve capacity is not measurable independently of the subsidiary task—it is in fact operationally defined in terms of it—it is obviously not possible, as Brown (1964) points out, to know whether scores on the subsidiary task are linearly related to any notion of 'reserve capacity'. Again, primary tasks will obviously differ in their amount of perceptual, motor and cognitive demand and, as Macdonald and Cameron (1973) point out, the concept of perceptual-motor load is not unidimensional; thus, it is hard to see how an index of reserve capacity is strictly comparable across different primary tasks. Further, adding a secondary task to the primary task changes the situation in which the latter is undertaken. The driver (or other subject) is, by definition, suddenly 'overloaded' to use up his total 'mental capacity'; he may find this stressful; and any increases in 'arousal' may alter his 'reserve capacity', making performance changes impossible to interpret. Finally, although the subject may be instructed not to let the secondary task interfere with the primary, he may well find it difficult to adhere to this instruction, and the primary task may well suffer in ways that are not obvious to an observer: for example, through less time and effort being devoted to planning and anticipating events and other functions which do not necessarily affect overt behaviour.

Is it possible to reliably discriminate different levels of driving task demand? If so, what are the effects of prolonged driving on the driver's capacity to deal with these demands? (Other 'fatiguing' conditions have not been investigated). Table 4.2 outlines some studies pertinent to the first question, and Table 4.3 some studies directed at the question of 'fatigue'.

Table 4.2.   Studies of subsidiary task performance during different levels of driving task demand.

Investigator(s)	Subsidiary task	Results
Brown (1962 a, b)	Identifying a changed digit in an otherwise identical repeating series	More errors in a busy shopping area than a quiet residential area
Brown and Poulton (1961)	Mental addition of three digits	As above
Senders et al. (1966, 1967)	Obscuring the driver's vision for different periods or at different speeds	Strong relationship between speed and occlusion time, but no effect of curves or other road features
Macdonald and Cameron (1973)	Press footswitch (one of two) to a tone presented to left or right ear	Response times followed a similar pattern to driving difficulty index in some cases
Curry et al. (1975)	(a) As in Brown (1962 a, b)	No significant difference between four traffic areas. When extremes of traffic compared dense traffic produced more errors
	(b) Random number generation	No effect of traffic areas
	(c) 'Verbal tapping': S required to say 'Tu' at regular intervals of c. 1 s	Greater variability in tap rate in dense traffic

Table 4.3.   Studies of subsidiary task performance and prolonged driving.

Investigator(s)	Subsidiary task	Experimental conditions	Results
Brown (1962 a, b, 1965)	(a) Attention task: S to detect odd-even-odd sequence in series of digits (b) Memory task: identify repeated letter in a sequence of eight	Performance (during driving) compared between test just after 8-hour driving shift with test just before 8 hour shift. Both tests at 4.00 p.m. (police drivers)	(a) Performance significantly better after shift than before (b) No significant differences
Brown et al. (1966)	'Random generation': S to say a month of the year at 'random' to the beat of a metronome (paced task)	Six performance tests at intervals throughout 12 hours interspersed with continuous driving (experimental day) or normal work (control day)	Performance consistently (but statistically NS) better on experimental than control day. No effect of time driving
Brown et al. (1967)	Interval production task—'verbal tapping'	As above	Higher variability (worse performance) on experimental than control day. No effect of time driving

While some subsidiary tasks do seem to be able to discriminate between very different traffic conditions, none of them seem to be sensitive to small changes in task demand. In some cases performance on the subsidiary task has not been significantly different between sitting in a stationary car and driving (e.g., the 'verbal tapping' task of Curry et al. 1975). A slightly contrasting approach to the measurement of the information load of a highway has been proposed by Senders et al. (1966, 1967). Rather than use up the driver's spare capacity by a subsidiary task, they prevented him from seeing the road for short periods; the driver either adjusted his speed according to the amount of looking time he was allowed, or adjusted the length of the visual occlusion according to his speed. A strong relationship was demonstrated between speed and occlusion times; however, other aspects of the road situation which appear relevant to its information rate (e.g., curves) had no effect on occlusion times.

The studies of prolonged driving in Table 4.3 show absolutely no impairment of performance; prolonged driving seems if anything to be associated with better performance. There is no obvious explanation why the 'random generation' task in the 1966 study was performed better during the prolonged driving day, while the interval production task was performed better on the normal working day. Brown's initial explanation of his early results is that they could have reflected fatigue due to other pre-shift activities; however, the later results he more plausibly explains in terms of the greater 'automatization' of the driving task during prolonged periods of driving, resulting in a greater 'reverse capacity'.

Brown in his 'fatigue' work has obviously conceptualized the problem of prolonged driving in terms of the task becoming an increasing mental strain on the driver, who is left with a smaller and smaller amount of reserve capacity (though his results have shown the opposite). A diametrically opposed point of view is contained in McBain's (1970) paper, which is concerned with the monotony of much commercial driving along the super-highways of America. It is ironic, as Harris *et al.* (1972; p. 46) point out, that roughly what McBain suggests as a means of alleviating boredom and increasing alertness has been used by Brown and his colleagues as a means of assessing the driver's decreased alertness and reduced mental capacity. McBain found that subjects who tended to have a higher rate of error on a very monotonous laboratory task also had a greater history of accidents, suggesting that boredom and monotonous conditions may well for some drivers reduce the safety of their driving.

The contrasting approaches of McBain and Brown might lead one to think that they are talking about two very different activities. Admittedly most of Brown's tests were carried out on busy city streets, while McBain's point of reference is the long, relatively featureless and traffic-free multi-lane rural highway; however, they illustrate the danger of building fairly elaborate theoretical models or experimental programmes on a few bland *a priori* assumptions about the task demands of driving. It certainly seems to be the case that the conditions under which Brown's drivers were operating were not sufficiently taxing as to put a strain on their 'mental capacity' (in fact the amount of actual driving done by the police drivers during their 8-hour shifts in Brown's early study was probably very small (Brown 1965)). However, the need remains for a much more sophisticated description of the driver's task under differing conditions, comparing, for example, the very different demands of the task of the city taxi and bus driver, the long-distance truck driver or coach driver, the commercial traveller and the holiday driver. Such a description would have to include both the type of roads and traffic conditions typically found by these drivers, as well as the constraints of schedules and deadlines they might have to fulfil, and non-driving aspects of their jobs.

Several studies do provide a basis upon which a comprehensive description of the task and job demands confronting drivers can be begun. Böcher (1975) has undertaken a very thorough analysis of the job of the truck driver in FR Germany. Edmondson and Oldman (1974) report that the drivers they interviewed found both monotonous driving conditions and congested traffic to be fatiguing; bad weather conditions were also considered to be taxing, and drowsiness associated with working night-shifts was an almost universal problem (see Section 2.8, p. 56, for a fuller discussion of this study). Concerning non-professional drivers, Tilley *et al.* (1973), from questionnaire responses of 1500 driving license-renewal applicants, related the experience of drowsiness while driving to driving habits and personal characteristics; one result in particular suggested to these researchers that certain characteristics of the driving task might be associated with drowsiness. Sixty-four per cent of their questionnaire population reported drowsiness while driving, while only 32% said they experienced drowsiness in other situations—and this they tentatively suggest may be due to stimulus deprivation during driving. Mellinger (1970) investigated the habits of drivers in planning and carrying out long driving trips; he found that while the majority of drivers interviewed planned and executed their trips allowing ample time margins, not over-reacting to delays but driving well within road and traffic conditions, and approximating their driving time to a normal day's activity cycle, a minority (about 10%) regularly over-extended themselves in the sense of driving long hours, behaving at variance with external limiting conditions, operating at near capacity, and at variance with their normal cycle of daily activities.

The absolute levels of task demand, specifiable independently of the driver, are perhaps not so important in terms of his mental capacity for the task as the way the driver paces himself during his driving so that he is never placed under a strain. This pacing is both a matter of planning and scheduling the trip beforehand, and of flexibly adapting himself to road, weather and traffic conditions. Mellinger's study suggests that there is a significant minority of non-professional drivers who find it very difficult to carry out such self-pacing effectively. It is quite likely that too-rigid scheduling of commercial vehicle trips, either by transport managers or by legislation (through the introduction of tachographs, for example), may in a similar way make it more difficult for commercial drivers to distribute their mental effort to the task in the most appropriate and least stressful manner. It is certain that the relative flexibility that the truck driver has in the pacing of his work is one of the most attractive features of his job. How this pacing works, how effective it is, and what happens when a conflict arises between the driver's optimal pattern for the allocation of effort and the requirement of the schedule he is working deserves much fuller investigation.

In conclusion, therefore, it is plain that the subsidiary task technique is not very effective in differentiating other than very gross differences in road and traffic conditions, nor in providing a sensitive monitor of any adverse effects of prolonged driving. Such improvements as have been noted after prolonged driving might be made possible by the increased 'automatization' of control skills as driving continues. Sufficient attention has not been paid to the wide variety of conditions that drivers confront, and the different levels and nature of task demand that these represent. Both congested traffic and monotonous roads are reported by drivers to be taxing, as are bad weather and night driving. The driver's ability to plan out his trip and pace himself throughout the driving period is probably crucial to his being able to avoid 'over-loading' or 'over-extending' himself.

4.6. Concluding remarks

Considering the number and scope of the studies discussed in this chapter, the extent of the conclusions that can be drawn can only be considered very disappointing. Even when giving maximum credence to those results which have shown evidence of some deterioration in performance, the most that can be concluded is as follows: there is very indirect evidence that prolonged driving and sleep deprivation may adversely affect the driver's peripheral visual efficiency, which may in turn affect his perception of speed; prolonged driving, particularly over several days and during irregular schedules, may be associated with a decline in steering precision and accuracy, and lane drift incidents may be more common during the night time hours, and towards the end of a week of irregular working hours; complex manoeuvring skills may suffer after prolonged driving; under certain circumstances (prolonged monotonous driving at night) a driver's ability to detect certain types of signals may be impaired; and drivers may tend to take greater risks in overtaking and generally interact with other road users in a less safe and courteous manner after many hours of driving. Very few of these positive results have been corroborated in more than one independent study and it would seem reasonable to conclude that, overall, there is little substantial experimental evidence that long periods of driving during day or night are associated with changes in driving performance involving increased risk.

Nevertheless it would be foolish, for three reasons, to infer from this conclusion that this is all there is to be said about 'fatigue' and driving. Firstly, professional drivers do complain about fatigue and drowsiness, and they report that they find certain conditions difficult and taxing (e.g., night driving and bad weather). Secondly, drivers have been known to fall asleep at the wheel even under experimental conditions. And thirdly, accident statistics do suggest that fatigue and drowsiness are small but significant contributors to HGV and other vehicle accidents.

Why then are these phenomena not represented in experimental evidence documenting the conditions under which driving skill has been shown to deteriorate?

A variety of reasons can be invoked to account for this. Firstly, the vast majority of studies have been one-off experiments involving the subjects in one driving session only; in such a design the social influence of the experimental situation is maximized and may well counteract any process in the 'fatigue' complex that may otherwise be evident.

Secondly, the allocation of research effort to different aspects of the driving task has been unbalanced. A large proportion of 'fatigue' studies have concerned the more rudimentary aspects of driving skill—the driver's physical control movements. It is these components of driving skill that would be expected to be, for most drivers, overlearned, highly automatic in their execution, and the least likely of all aspects to be susceptible to deterioration under adverse conditions. In another large group of 'fatigue' studies—and the 'vigilance' studies are the most obvious examples—an experimental methodology and theoretical framework developed under very different circumstances has been tagged on to driving research with very little consideration of its relevance to that skill or its interference or interaction with it. In contrast, very little experimental effort has been devoted to more complex aspects of driving—decision-making, risk-taking and processes governing attention. It is these aspects of driving that are likely to be most affected, particularly if one places much emphasis on the motivational aspects of 'fatigue'.

A third reason concerns the concept of 'fatigue' that has achieved hegemony throughout the bulk of this research. As Laurell and Lisper (1976) point out, the goal of most 'fatigue' research has been to develop a reliable measure of 'fatigue'; and this notwithstanding Muscio's (1921) strictures on this very matter. Thus the major emphasis has been on selecting suitable dependent measures, often with scant regard for their actual significance to the driving task as a whole, but which will simply show some change over time that can be attributed to 'fatigue'. 'Fatigue' has thus acquired the quite unjustified status of a causal agent, a hypothetical mediating state of the organism, but one which has never been precisely defined nor the laws of its movement articulated. This has had several unfortunate consequences on the nature of the research that has been conducted, bearing on the antecedents of 'fatigue', the nature of the skill that is supposedly affected, and the way in which the 'fatigue' process operates.

With the major exception of Human Factors Research, Inc., there has tended to be an assumption that the number of hours spent sitting in the driving seat is the only independent variable of interest. It is likely that if the driver's capability to perform his task adequately is impaired, this will be the result of the interaction of a multiplicity of factors which will include, of course, how long he has been driving and what rest breaks he has had, but also the time of day, what sleep he has had and when, the cab environment, the road, traffic and weather conditions, and his recent social interactions at home and at work. There has also been an implicit assumption that the task of driving does not vary very much and that, again, if one has specified its duration, that is sufficient. Driving can be, at times, monotonous, physically undemanding, and visually and mentally unstimulating. At other times the driver may be taxed by the demands of the task. The problems of a truck driver on the night shift on clear empty multi-lane highways are very different to those of another driver who is delivering goods to a series of congested city depots to a tight schedule during the day, yet the complaints of both may well come under the heading of 'fatigue'.

Thus the typical 'fatigue and driving' experiment confounds a whole range of environmental and task-related factors in an experimental design that counterposes hours of driving against an average performance score for a group of subjects, usually non-representative of the driving population. Results are searched for some evidence of a linear decrement in performance as a function of time driving, the rationale presumably being that if a detectable change in performance is obtained after x hours of driving, one can then extrapolate to $x + n$ hours when the change in performance has become so magnified as to be totally incompatible with safe driving. It is as if a small decline in fine steering reversals between the fourth and the sixth hour of driving could be taken to show that by the eighth hour the driver will be unable to keep the car on the road. The ambiguity of the phrase 'significant decrement in performance' further adds to the confusion, creating the impression that some minuscule, statistically significant result has some significance for road safety. Critical incidents that have arisen in the experiment, where some near-accident may have occurred, for example, are treated in anecdotal fashion as if these extreme cases, where the driver's skill really has failed, are an embarrassing aberration from the regularity of predicted experimental results. Research into the deterioration of driving skill should start with the analysis of critical incidents which can point to the way in which skill has deteriorated when the driver has suffered from fatigue, and the circumstances under which this tends to occur. Unfortunately research into the breakdown of driving skill has almost totally eschewed this approach.

It is of course possible to argue that it is misguided to approach the fatigue problem by looking for deterioration in driving skill. For most of the driving population, and especially professional drivers, the skill is well practised and learned; under most circumstances the general capability to recognize and compensate for fatigue or drowsiness may mean that driving will be continued at a high level of proficiency until the driver is in a state of exhaustion and near collapse, when he may suddenly cease to function as a driver (by falling asleep for example). In such a case the question of the deterioration of driving skill is not very important. Driving can be regarded as an activity, continued performance of which *per se* is not sufficient to cause a deterioration in its practice. Rather, it is when continued driving interferes with needs like sleep, or food, or varied sensory stimulation that the driver begins to function at a reduced level. However, because the demands of driving are not that great, he will continue to be able to perform adequately until a crisis point occurs when he suddenly collapses and is unable to continue. There may well be some truth in such an argument, but particularly if one considers driving skill only in terms of the drivers' movements of the vehicle controls. (Though, sudden blackouts do occur and certain people are apparently susceptible to rapidly and briefly falling asleep without warning; Grubb (1970) and Ford (1971) discuss this problem of narcolepsy and driving.) However, driving is frequently continued for extended periods where the driver is performing at a reduced level of skill, where he is attending only inadequately to signals and events in his environment, and where his ability, or motivation, to make safe decisions may be adversely affected. It is these processes of increasing deterioration and disintegration of skill that remain the proper object of 'fatigue' and driving research.

Thus to conclude, although the majority of studies of driving performance and 'fatigue' have tended to focus on the perceptual-motor skills involved in tracking, and notwithstanding the limited success of measures of steering performance, an explanation for those accidents which are attributable to 'fatigue' should not be sought solely in terms of the driver's manual or pedal dexterity. It is likely that many 'driver-falling-asleep' accidents are preceded by an increasing deterioration in the way the driver is paying attention to his vehicle and to the road environment. There is a certain amount of experimental evidence for gross disturbances in the quality of the driver's attention (see Section 2.8 and Chapter 6). Only a tentative start at a more scientific account of these processes is now possible however. It is also possible that exposure to many of the stresses in the driver's physical and social environment may make for a deterioration in other aspects of driving skill, particularly his powers of decision-making and his propensity to take risks, as well as to the quality of his attention. Such changes may make a contribution to a number of accidents (it is impossible to guess how many) which are not directly or overwhelmingly attributable to fatigue. There is a small amount of empirical evidence to support this, and a theoretical explanation of these processes is probably best accomplished in terms of motivation. These conclusions are necessarily very tentative; a complete account of the processes involved in the deterioration of driving skill has hardly begun.

Chapter 5
The Physiological Status of the Driver

5.1. Introduction

If drivers are functioning less efficiently after a number of hours of driving, when working certain shifts, or driving at certain times of the day, it would be expected that physiological recordings would detect significant changes in activity; though to return briefly to Bartley and Chute (1947), there is no logical connection between physiological, behavioural and subjective states. The most closely related physiological state to the subjective condition of fatigue is what these authors call 'impairment'. The criterion of 'fatigue' in this conceptual scheme is generally a strong desire or need to terminate or escape from a particular situation. 'Impairment' implies some measurable degradation in the physical state of the organism. In what has come to be recognized as one of the most important aversive situations which typify the truck driver's work—prolonged periods of monotonous, low-incident driving, combined with various shift systems, and/or periods away from home—the terms 'impairment' and 'fatigue' take on a slightly different meaning from their normal association with heavy physical labour. Many drivers do, of course, have the hard physical work of loading and unloading their vehicles; however, particularly for long-distance drivers, this is not usually a major part of their task.

Within this type of situation, characterized by a low level of stimulation, the typical physiological and subjective syndrome that has been the model for the majority of studies has been of a decreasing level of responsiveness in the driver; physiological activity declining towards a level characteristic of extreme inactivity, drowsiness or sleep; and strong feelings of boredom and monotony, tending towards a trance-like state where misperceptions and maybe hallucinations are not uncommon. The interpretive framework that has fitted this model best and has with few

137

exceptions been applied in its simplest form has been the 'activation hypothesis' of Lindsley (1951), Duffy (1957, 1962) and Malmo (1959). The hypothesis is 'that intensity is a characteristic of behaviour which can be abstracted and studied separately' and further, in Lacey's (1967) words, that there is 'a unidimensional continuum of arousal ranging from coma to the most excited and disorganized forms of stressful behaviour'. Degree of excitation, arousal, activation, energy mobilization and sometimes alertness are all taken to be equivalent terms.

The confounding of behavioural, physiological and subjective events is designed into this theory and is one reason, together with its considerable face validity, why it has proved extremely popular as an umbrella theory which covers a wide range of phenomena. It has been criticized by Lacey (1967) on several counts, which can best be interpreted as making clear the severe limitations of the theory rather than saying that the theory itself is wrong. The first of these criticisms is that it is possible to conceptually distinguish and empirically dissociate different forms of arousal—autonomic, electrocortical and behavioural. Secondly, correlations among various measures of arousal tend to be very low. It is accepted by activation theorists that interindividual differences in arousal measures will be high; however, the theory is based on the idea that "individuals exhibit ... idiosyncratic but highly stereotyped patterns of somatic and autonomic activation" (Schnore 1959). Lacey contends that these intra-individual correlations in arousal indices are also too low, and that correspondence among physiological measures is only in the direction of change rather than in the amount of change. Finally, as well as individuals exhibiting characteristic patterns of response, different situations also elicit particular response patterns (a process he labels 'situational stereotypy'). To exemplify this, Lacey points to the classic work of Ax (1953) and others in the differentiation of different types of anger, fear, anxiety, etc., as well as to his own work on different types of task performance.

Lacey has developed a hypothesis concerning the importance of one autonomic component—the cardiac system—in monitoring or mediating perceptual inflow to the brain. It states 'that "mental concentration" is accompanied by cardiac acceleration, and that attention to the environment is mediated by cardiac deceleration' (Lacey *et al.* 1963). Cardiac deceleration has been found in a large number of situations like looking at pictures, searching for hidden figures, and in the foreperiod in a reaction-time sequence, while an acceleration has been found in a variety of cognitive tasks. The instrumentality of the heart rate response is open to question (Hahn 1973). Fuller *et al.* (1975) failed to obtain consistent results in agreement with this hypothesis when heart-rate levels were manipulated by the inhalation of nicotine. The physiological basis of this hypothesis with the chain of processes, from baroreceptors in the aortic arch and carotid sinuses, to inhibitory effects on the ARAS (ascending reticular activating system) to the cortex, is particularly difficult to verify. However, the development of the concepts of directional fractionation and situational stereotypy have increased considerably the predictive power from stimulus situation to behaviour and physiological response.

Other developments from, and criticisms of activation theory will be considered below as they arise concerning particular physiological measures.

Much of the discussion will centre on the interpretation of physiological measures. Trends in these measures in the typical monotonous driving situation do not fit easily into a description in terms of 'impairment'. A criterion of when someone is at a dangerously low level of physiological activation or is, for example, about to fall asleep, is not simple to prescribe; the use of feelings and mood as an aid in the interpretation of physiological studies has not yet reached a very sophisticated level (this is discussed in the next chapter). Again, hampering the straightforward interpretation of physiological measures is the problem of individual differences in activation levels; the results from nearly all the studies to be discussed below have been reported as pooled data for groups of subjects. This has meant that trends in the data have nearly always been of a small magnitude, frequently inconsistent and statistically insignificant; meaningful variation within individuals may well have been obscured.

The bulk of research has recorded some aspect of the cardiac system, particularly, of course, heart rate. A number of investigators have investigated changes in heart rate variability; some have looked at the ECG cycle, and a few have measured blood pressure. Other autonomic measures that are represented are respiration rate and skin conductance. One or two studies have measured levels of endocrine hormones related to the stress response. Cortical arousal has been measured by the EEG, and in a few studies by critical flicker fusion thresholds. Somatic measures have included EMG recorded from neck muscles, posture, eye blink rate and electro-oculograms; these measures presumably reflecting drowsiness or a degraded level of activity.

5.2. Heart rate

One of the most common findings with heart-rate (HR) measures has been an increase in cardiac frequency when driving has begun (e.g., Dupuis 1965, Dupuis and Hartung 1968, Hashimoto 1967, Wyss 1970, Lecret and Pottier 1971, Lecret and Niepold 1974). Driving under different road and traffic conditions typically produces different magnitudes of response. Helander (1976) summarizes the main findings from a number of studies as follows: "Generally, driving on urban highways without much traffic interference produces heart rate increases of the order of 10 bpm (beats per minute). Under more restricted conditions, (town driving, rural roads, driving on mountain roads) increases of 30 bpm are typical, and during traffic events of short duration (overtaking, stopping, etc.) transient increases of 45 bpm might be observed".

Hoffman and Schneider (1967) present their data from a series of investigations in a way that makes more clear the variability in response between drivers. The data are presented as percentages of the 600 healthy drivers in their studies who had certain percentage increases in heart rate over resting levels. Absolute levels are not given. No drivers during highway driving in low-density traffic produced more than a 20% increase; 28% of drivers during urban driving produced increases of 20% or more; and in 'critical situations' (overtaking, sudden stops, etc.) 42% of drivers produced a 20% increase. Forty per cent or greater increases were found in 8% of subjects during city driving and 14% of subjects during critical situations. Speed seemed to have no effect on heart-rate levels. Level of experience did have an influence with a lower proportion of experienced drivers showing large HR increases in urban traffic.

How should such heart-rate increases be interpreted? The two main interpretations have emphasized physical workload effects and emotional factors, though it is also likely that attentional factors play some role in heart-rate changes during driving.

Wyss (1970) emphasizes the physical work involved in maintaining body posture and balance; such energy expenditure is reflected in heart-rate differences between driving uphill and downhill and along flat roads. Helander (1976) reports a further study by Wyss which compared the O_2 pulse (the ratio of oxygen consumption and simultaneous heart rate) for driving along various types of road and riding a bicycle ergometer. Compared to the ergometer there were higher heart-rate values for equivalent levels of oxygen consumption in the driving situation. The interpretation offered for this is that emotional factors in the driving situation produce the increased heart-rate levels.

Another suggestion has been that the physical effort involved in the control movements involved in driving could account for the heart-rate increases (Rutley and Mace 1972). However, it seems unlikely that this is an adequate explanation in view of the following findings: changing from manual to automatic transmission results in no difference in heart-rate level (Seydal 1972); the heart-rate responses of drivers and passengers in a car have been found to follow an almost identical pattern during 30 minutes of city driving (Simonson *et al.* 1968), which these authors interpret as again reflecting the emotional stress of driving; and furthermore, heart-rate increases have been found where no control operations were necessary (Hoffman and Schneider 1967).

There is also evidence that some heart-rate changes may reflect the driver's attentional processes. Watts (1977) found a significant decrease in heart rate as a response to 'rumble strips' on the roadway (strips of coarse textured material to warn drivers of approaching roundabouts etc.). This response was most marked for subjects unfamiliar with the test site. Platt (1969) found heart rate to decrease when his subject drivers were passing oncoming cars; conversely, increases in heart rate were found at freeway entry ramps. Following Lacey's hypothesis (Lacey *et al.* 1965), one would expect that in situations of passive attention to the environment, like passing oncoming cars and hearing rumble strips, there would be cardiac deceleration, while in situations requiring decision-making and preparation for action (like freeway-merging), cardiac acceleration would be expected. 'Directional fractionation' occurs when the cardiac response, for example, is in the opposite direction to other physiological response measures. Helander (1976, 1978) provides an example of this where bumps in the road elicited GSR (galvanic skin response) peaks but no corresponding heart-rate increases. Helander concludes from his very detailed study of a small section of road that psycho-physiological response levels (GSR, HR) were positively associated with levels of task demand and the complexity of the road and traffic environment. Whereas heart rate was a good indicator of gross environmental characteristics, it was less rapidly responsive to minor fluctuations in task demand than (GSR).

Thus the evidence suggests that, over short periods of driving, changes in heart-rate levels reflect variation in at least the three factors of the maintenance of the body's balance and posture, the emotional stress of driving, particularly in traffic, and the attentional and cognitive demands of the task.

When HR has been a dependent measure in studies involving long periods of driving, the independent variable has most often been simply time spent driving. Many studies have not even specified the general road type, traffic conditions, weather, time of day and other similar parameters. Many of these factors could well have been changing throughout the routes chosen in these studies, with associated changes in HR and other physiological variables. An approach which has only rarely been attempted (e.g., by Sussman and Morris (1970) with GSR to emergency situations) has been to investigate physiological responsivity to particular types of stimulus situations. Such an approach could well throw up results more indicative of degradiation of physiological response than an analysis in terms of gross levels of activity. However, the studies will speak for themselves.

Table 5.1 summarizes the main findings of mean heart-rate changes from a number of studies which have involved driving periods of three hours or more. A number of different factors can be invoked to account for the observed variation in heart rate. These include circadian variation, adaptation of circadian rhythms to rotating shift schedules, the effects of prolonged driving, adaptation to the experimental situation, and the cardiac responses to physical work and to heat. It is not easy to separate out the independent effects of all these variables, for in many of the experimental situations a number of different effects are confounded.

There is a pronounced circadian variation in heart rate (Kleitman and Ramsaroop 1948) and it seems likely that a major part of the variation in heart rate found in these studies reflects this cyclic fluctuation. Only a few of the studies have involved driving at times which span the whole 24-hour cycle, and notable among these are those by Mackie and Miller (1978) and Harris *et al.* (1972), both of which studied professional drivers in a close approximation to their normal working circumstances. Thus, in the former study, regular daytime trips tended to be associated with an increasing heart rate, and trips which went from the late evening to the early morning hours were characterized by a falling heart rate. An overall circadian trend is also evident in Harris *et al.*'s study. The majority of the other studies in the table were conducted at night starting at any time after 7 or 8 p.m., and their results are congruent with the driving trip coinciding at least partially with the declining portion of the circadian rhythm. The only study which consistently found reliable increments in HR with time driving was that of Mackie *et al.* (1974), which was conducted during the day. Strong support

Table 5.1.   *Main findings of studies recording heart rate over prolonged periods of driving.*

Investigators	Experimental situation	Time of day	Length of drive (h)	Recorded changes in heart rate
Michaut and Pottier (1964)	Car, closed track	1415–1715	3	7 of 10 subjects showed decrease
Lecret and Pottier (1971)	Car, autoroute	(a) 2100–0100 (b) Night	4 6	(a) Second 2 h of 4 h drive always lower than 2 h drive (b) Decrease throughout trip
Lecret and Neipold (1974)	HGV, autoroute	Night	3	Decrease to just over resting levels
Dureman and Boden (1972)	Simulator	Day, unstated	4	Shock group: 12 bpm decrease; non-shock group: 11 bpm decrease
Lisper *et al.* (1973)	Car, closed track	Morning, start c. 0930	3	Experienced 15 bpm decrease; inexperienced 7 bpm decrease
Sugarman and Cozad (1972)	Car, highway	Night, start 2000	4	Mean 8 bpm decrease
O'Hanlon and Kelley (1974)	Van, highway	Night, start 1900	5	Up to 10 bpm decrease
Harris *et al.* (1972)	HGV, all roads	All times of day and night	Up to 10	Mean 10 bpm decrease
Mackie *et al.* (1974)	Car and HGV, highway	Day, 0800–2000	Up to 10	Hot condition: mean 20 bpm increase; comfortable temp: mean 6 bpm increase; high and low noise and vibration conditions, no significant time or condition related effects.
Yajima *et al.* (1976)	Car (a) highway (b) closed track	Day and night	(a) 7–10 (b) 24	No significant time related changes reported
Riemersma (1976)	Car, highway	Night, 2200–0600	8	Mean 10 bpm decrease
Mackie and Miller (1978)	HGV, bus	Day and night. Regular and irregular shifts over 6 days repeated	Up to 10 h, less on irregular shifts	Circadian trends most pronounced, but modified either up or down by other factors. Mean changes all less than 10–12 bpm

for the circadian hypothesis is provided by Riemersma (1976). He found that heart-rate levels recorded from subjects during driving at 6 a.m. did not differentiate between a group of subjects who had just had a night's sleep and the same subjects after eight hours of driving. As he comments, a lower heart rate could be interpreted in terms of 'de-arousal', a higher heart rate could be taken to reflect a greater effort to counteract sleep loss, but no difference would seem to suggest that heart-rate levels are not affected by eight hours of driving.

On the other hand Harris *et al.*'s (1972) study provides evidence that both time on task and diurnal variation have an effect on heart-rate levels. Figure 5.1 shows the overall diurnal variation in heart rate by hour of day. The data are from commercial vehicle drivers driving relay-type operations and recorded during 90 runs occurring throughout the 24-hour day. Thus each data point represents the average of a large number of recordings made after different periods of driving, and is expressed as a deviation from the grand mean of all recorded heart-rate periods. Figure 5.2 shows a

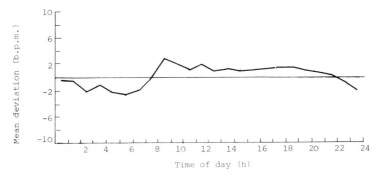

Figure 5.1. *Diurnal variation of heart rate of truck relay drivers. After Harris* et al. *(1972).*

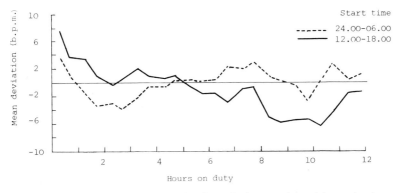

Figure 5.2. *Variation of heart rate of truck relay drivers by hours on duty: night starting time vs. day starting time. After Harris* et al. *(1972).*

similar presentation of data but as a function of time driving rather than time of day. These graphs are based upon two groups of drivers who had started their shift either between midnight and 6.00 a.m. (19 out of 21 before 3.00 a.m.) or between noon and 6.00 p.m. (again, 20 out of 23 before 3.00 p.m.). Harris and his associates identify the period from the eighth to the tenth hour of driving in particular as showing for both groups of drivers a lower mean heart rate than would be expected on the basis of diurnal variation alone. However, their analysis is not carried beyond this descriptive level.

There were also deviations from the circadian rule in Mackie and Miller's study. Thus the increase in heart rates during the final daytime driving trip of the week was frequently rather lower than it had been earlier in the week. Occasionally this happened at other times, and sometimes there was a higher than expected average level of heart rate during the early morning hours (particularly at the end of trips). Mackie and Miller tend to interpret the former effect as evidence of fatigue, producing 'de-arousal', and the latter as reflecting increased effort to counteract fatigue.

However plausible this interpretation might be, it would be foolish not to be aware of the logical danger of assuming what one wants to prove—the assumption being that any deviation from regularity reflects fatigue. It is very difficult to evaluate the validity of these interpretations because potentially corroborating trends in other variables are correlated neither systematically nor statistically (though they are mentioned in passing where they support the argument), and because there is absolutely no treatment of individual differences in response which might point to occasions and individuals where a systematic and serious deviation from the norm might have happened. There is an alternative explanation which could possibly account for some of this apparent flattening of the circadian rhythm when it occurs towards the end of a week of irregular driving shifts. One would expect that the rotating shift routine experienced by these drivers would have some effect on the circadian heart-rate rhythm, causing it perhaps to flatten, to shift phase, or to be otherwise disrupted. It is difficult to separate out these potential effects of circadian disturbance from those which might more appropriately be described as due to fatigue following prolonged work.

It should of course be emphasized that whether the heart-rate changes are related to circadian or time on task factors does not affect whether we describe such changes in 'arousal' terms. Indeed, the fact that Kleitman and Kleitman (1953) did find that heart-rate periodicities fairly accurately followed an altered period of sleep, eating and daily activity patterns, might suggest that it is a more accurate reflector of 'activation' than a more endogonous rhythm like body temperature.

Some of Harris *et al.*'s (1972) findings do lend credence to an interpretation of at least some drivers' heart-rate changes in terms of fatigue or de-arousal. Seven drivers were reported by the experimental observers as being 'fatigued' (they "looked tired, acted tired, were irritable" etc.); these drivers had a more consistent and larger drop in heart rate during their runs (an average of nearly 19 bpm), as well as having a lower average heart rate than other drivers. Similarly, drivers who showed a greater rate of driving errors (particularly lane drifting) showed lower levels and steeper declines in heart rate. Similar results were also obtained for sleeper drivers who were at the end of a cycle of runs compared to drivers at the start of a cycle, suggesting a decrease beyond the expected circadian variation.

There are three studies which run counter to the predicted direction of nychthemeral change. Dureman and Boden (1972) do not state the time at which their experiment took place but it seems likely it was within normal working hours. Lisper *et al.*'s (1973) study was conducted in the morning. Michaut and Pottier (1964) had their subjects drive in the early afternoon, which is probably too early in the day to predict a declining circadian phase for most subjects. All of these studies administered either auditory reaction time or vigilance tests during the driving session. In Dureman and Boden's experiment one group received a small electric shock contingent upon driving errors. Another group had no shock but heard a click instead. Quite apart from the electric shock, the knowledge of results provided by the click and the presentation of the reaction time and vigilance tasks would be expected to be arousing for the subjects. One can only speculate as to whether these factors made for an abnormally high heart rate at the beginning of the experimental period.

It is possible to develop this last point to cover virtually all the studies reported. Most of the reported decrements in heart rate have been of the order of 10 bpm, which is not a very large change; the evidence reviewed at the beginning of this section suggested that heart-rate increases of up to 10 bpm above resting levels would be expected on highways without much traffic interference, and a proportion of this increase could be attributable to emotional factors. The arousing effect of an experimental situation, with instrumented vehicles, electrodes to be attached, the feeling that performance is being assessed, etc., might well show itself in initially highly elevated heart rate levels (Zwaga (1973) provides a good discussion of some of these problems). In many of these studies the manner of data presentation obscures when during the driving period the major decrements in HR have occurred. However, in those studies where it is possible to tell, almost without exception the bulk of the decrement has occurred fairly early on in the driving period. Dupuis (1965) reports HR declining to just above resting levels in less than 20 minutes among his subjects. In other more prolonged studies the bulk of the decrement has occurred within the first 2–3 hours and frequently much less, with very small changes throughout the rest of the trip—up to 8 or 10 hours in duration. This is certainly true of Sugarman and Cozad (1972), Dureman and Boden (1972) (despite the brevity of their experimental periods) as well as Harris *et al.* (1972) and Riemersma (1976). Considering such factors as these it is not hard to concur with Riemersma's (1976) suggestion that the declining heart rate represents not so much 'fatigue' or significant de-arousal as adaptation to the task (decreasing stress, habituation to sensory stimuli, etc.).

Part of the difficulty in interpreting physiological studies is that most often there is no criterion against which the results may be judged. If one takes a 'resting level' as the norm and emphasizes the stressful aspects of the driving situation, then one is inclined to see a decline towards that level as adaptation. It is as easy to take a contrasting view that would emphasize that any deviation from the initial 'stabilized' levels represents some kind of falling off from a particular level of physiological functioning which is the optimum for the performance of that task. Not only is the cardiovascular system very complex in that changes in heart rate could reflect a whole variety of different factors, but there are also great individual differences not only in the degree of response but also in the manner of responding. Cullen (1976), for example, has reported that some individuals respond to a tilt test with changes in cardiac frequency; with others the major response is a change in blood pressure. These differences seem to be related to measurable personality factors.

Two rather different factors have also been shown to have a marked effect on HR levels while driving. Mackie *et al.* (1974) found that in a hot car and truck cab (temperature at either 80° or 90° F, WBGT) there was a larger increase in body temperature than in a comfortable environment. There was no effect of noise or vibration levels. The effect of a moderately heavy loading task (itself producing a higher heart rate) was found by Mackie and Miller (1978) to persist throughout subsequent driving periods, making for higher average heart rates than during a week of driving when no loading had been done. This effect was most marked at the beginning of the week, suggesting a gradual adaptation to the loading task among the drivers, who were all unused to performing such work as part of their normal job.

Thus research into variations in heart rate during prolonged driving is subject to frequently contradictory interpretation. The evidence favours circadian rhythmicity as being the strongest influence on such variation, though fatigue related to prolonged driving may modify this—the evidence for this is clearest when the circadian trend is exacerbated, particularly towards a reduction in heart rate, though increases in heart rate (or a flattening of the circadian trend) may also reflect effort to counteract fatigue (for example, at the end of an early morning trip). However, such effects may be confounded with adaptation of the circadian rhythm to the changing shift pattern, and/or the emotional adaptation of the individual to the unusual experimental situation.

5.3. *Heart-rate variability*

Recently, a number of researchers have put forward the idea of heart-rate variability (HRV) being an indicant of alertness or arousal. Much of this work applied to driving has been done at Human Factors Research, Inc. in California (e.g., O'Hanlon 1971). In a resting or relaxed state the momentary heart rate is not constant but tends to fluctuate, and this fluctuation can be of up to 10–15 beats per minute (Kalsbeek and Sykes 1967, Kalsbeek 1971). When a subject is doing dynamic or static physical work, both the general level of heart rate rises and the extent of the irregularity decreases. However, during performance of a perceptual or mental task it has been found that the general level of heart rate will stay approximately the same but the irregularity decreases. Thus it would seem that the irregularity of heart rate and the change in overall level in heart rate are independent phenomena (Kalsbeek and Ettema 1963). Kalsbeek and his associates have suggested that HRV or sinus arrhythmia is a fairly sensitive index of mental load and can distinguish different levels of performance. Typically, binary choice tasks have been used (Kalsbeek 1973). However, Obrist *et al.* (1964) were not so successful in relating HRV to performance levels on pattern-recognition and mirror-tracing tasks; there was no correlation between HRV and perceptual task performance, while speed on the sensorimotor task tended to be positively associated with greater amplitude of HRV. However, HRV was lower during performance of each task compared to resting levels.

Research which has recorded some measure of heart-rate variability over prolonged periods of driving has not, generally, obtained the same consistency or magnitude of results as have been found with measures of heart rate. The series of research programmes incorporating HRV measures by Human Factors Research, Inc. originated with O'Hanlon's (1971) report. HRV was recorded during runs in a van which took place throughout the 24-hour cycle. Results were presented only descriptively without any inferential statistical treatment. The general findings were that HRV tended to increase with driving time (this was true for every run); there was a 'recovery' decrease in HRV after rest breaks; and critical events during driving were characterized by a reduction in HRV. Neither Harris *et al.* (1972) nor Mackie and Miller (1978) found any consistent and reliable trends in HRV with their professional drivers. Some groups of drivers seemed to show some upward trends in HRV, but others did not, and rest breaks had no consistent effects. Further, there were some findings which are not readily explicable in terms of the theoretical foundation for this work—some older drivers, for example, somewhat later on in runs had very low heart-rate levels as well as very low HRV scores. O'Hanlon and Kelley (1974) had more success, finding that the mean coefficient of HRV rose from the first to the last driving segments in each of their series of experiments; but, again, this conclusion was not based on statistical inference but a descriptive analysis. Mackie *et al.* (1974) found no significant changes in HRV over time under different levels of noise and vibration. In some cases, though not always, driving in a hot cab or car was associated with higher levels of HRV, and increasing HRV with time, which seemed to relate to subjective reports of drowsiness. However, in other similar conditions this result was not replicated.

Researchers from other institutions have also recorded HRV. Sugarman and Cozad (1972) found slight, significant increases in HRV over prolonged driving, but found this measure the least sensitive of the psycho-physiological measures they used. Riemersma (1976) obtained very similar results, and further found that HRV levels did not discriminate between the very different levels of traffic density between the pre- and post-test driving situations.

There are several reasons which might account for this not very inspiring catalogue of results. Firth (1973) has identified three main problem areas in research on cardiac arrhythmia and mental load, which apply with the same force to research on specific tasks like driving. These relate to the definition of mental load, the many factors that affect HRV, and the variety of ways of expressing HRV in numerical terms.

Regarding the first, definitional, problem of heart-rate variability as a measure of mental load or as an alertness indicator, there is no agreed definition of mental load; it is not clear that any useful unified concept of mental load is possible. It is likely that short-term changes in cardiovascular response are related to qualitative aspects of the stimulus situation, the response demands of the situation and the outcome of the response. Such aspects are not easily integrated into a quantitative mental load construct, quite apart from the frequent difficulty in deciding what the subject is responding to. Thus it is not surprising that different studies have found conflicting results. For example, Auffret *et al.* (1967) found that in pilots, differential information-processing demands of two types of cockpit display console during landing manoeuvres were reflected in HRV measures; on the other hand Pin *et al.* (1969) found that in drivers, different road and traffic conditions did not produce HRV differences. In the studies of prolonged driving discussed above, the level of mental load was not manipulated, and presumably the demands of the task remained the same throughout the drive. Thus the level of HRV is presumed to reflect the amount of attention allocated to the task at any time during the trip, and further HRV becomes an alertness indicator (O'Hanlon 1971), because attention and alertness both belong under the semantic umbrella that is the arousal hypothesis. There are two difficulties with his formulation, one logical, the other conceptual. Assuming that HRV does show measurable differences between levels of task difficulty, it does not follow necessarily that a change in HRV will imply a change in task performance. And there is little evidence that HRV is a sensitive measure of general activation; Volow and Erwin (1973), for example, failed to find any significant relationship between HRV and drowsiness onset.

If HRV is related to mental load, it is also affected by many other factors, from dynamic and static work-load to emotion and age (Firth 1973).

Sayers (1973) found that the bulk of the variability in heart rate could be accounted for by oscillations in the blood pressure dynamic control system and the respiration cycle, and suggests that HRVs associated with mental load are probably mediated by and secondary to these respiratory and pressure vasomotor changes. Both blood pressure and respiratory measures have been recorded over long driving periods. However, it is not always easy to relate these measures to HRV measures even when they have been undertaken in the same study. Mackie *et al.* (1974) recorded oxygen consumption and ventilation volume from truck and car drivers. Truck driving was associated with small but significant increases in these measures; this is most likely a result of vibration. In this experiment there were no significant changes in HRV or blood pressure. Both Dureman and Boden (1972) and Michaut and Pottier (1964) found significant decreases in respiration rate during their experimental sessions. The decrease reported by Lisper *et al.* (1973) was not statistically significant.

Yajima *et al.* (1976) report a decline in 'inspiratory volume and number', though they presented no data on this. It seems reasonable to suggest that in a vehicle with a smooth ride, such as a car, van or simulator, respiration rate may tend to decline; increases in HRV may also occur associated with this. Under the hyperventilatory effects of vibration (as in a truck) one would not expect a decline in either measure.

Blood pressure measures do not lend themselves to a straightforward interpret-ation. Michaut and Pottier (1964) report a significant decrease over time; Yajima *et al.* report a decrement in pulse pressure (maximum minus minimum blood pressure); Mackie *et al.* (1974) report no consistent time-related changes except in one condition of high temperature, where, as predicted, systolic blood pressure was lower than in comfort and decreased throughout the trip (this was not matched by another heat condition where blood pressure was higher for some, possibly stress-related, reason). The directions of these changes are not obviously related to HRV changes.

Heart-rate variability is not entirely independent of heart rate; Opmeer's (1973) analysis showed that heart rate accounted for about 50% of the variance of a complex of measures (the 'D complex') of HRV. Related to this, Firth (1973) invokes general physiological principles like the 'law of initial values' and the 'law of autonomic balance' (which suggest certain optimum limits for physiological functions), and he postulates that at higher levels of heart rate, heart-rate variability would be diminished. This is particularly relevant under extreme stress or fatigue, probably less so under most driving situations.

Firth's third problem area is the quantification of HRV. Different methods will give different scores, and there is no standard procedure. With minor variations of detail, there were two or three basic scoring techniques used in these studies of prolonged driving. O'Hanlon (1971) and Sugarman and Cozad (1972) defined HRV as the average deviation of *n* inter-beat intervals, updated with each successive interpulse interval. Harris *et al.* (1972), O'Hanlon and Kelley (1974) and Mackie *et al.* (1974) calculated the standard deviation of instantaneous heart rates (for each inter-beat interval) over a period of either 30 seconds or 1 minute. A coefficient of variability was derived by dividing this standard deviation by the mean heart rate for the period. Sometimes results are presented as a standard deviation, sometimes as a coefficient, and sometimes it is impossible to say which; no rationale is presented. Riemersma (1976) employed the standard deviation of inter-beat intervals over a two-kilometre road section as his measure of HRV. Some confidence that these different techniques are measuring the same thing is provided by Opmeer (1973). He performed a factor analysis and frequency analysis of 22 different scoring methods and found that scores based on the summation of successive differences in heart rate and score based on standard deviations covaried highly together; he called this collection of scoring methods the 'D complex'. Egelund (1982), however, has suggested that power in the band around 0·1 Hz of spectrally analysed HRV is the most sensitive indicator of fatigue and the least susceptible to the influences of physical workload and the random variation of heart rate. But the criteria of sensitivity to fatigue he used were not particularly arduous or comprehensive—a correlation with time driving during a 4-hour drive in the afternoon, and agreement with the drivers' post-driving reports of fatigue.

5.4. Changes within the ECG cycle

Simonson (1961) provides a wealth of clinical evidence that depression or inversion of the T-wave in the cardiac cycle is a reliable response to stress. He and his colleagues have investigated the ECG of drivers both during short periods of rush-hour driving in a car, and during more prolonged driving periods (Burns *et al.* 1966, Simonson *et al.* 1968). One subject was studied during eight 30-minute rush-hour drives over four days. During each drive there was, in the authors' words, "a distinct lowering or flattening of the T-wave, quite similar on the different days" (Simonson *et al.* 1968; p. 131). Results from four healthy subjects who drove over prolonged periods were summarized thus:

In C the T-wave decreased after four hours of driving to half its original size. In A, transient flattening of the previously positive T-wave developed after four hours of driving with some correlation to road events, and in B transient inversion developed after one hour. As driving continued the periods of T-wave flattening in A or inversion in B became more frequent and more prolonged. After interruption of driving, the T-wave recovered to approximately two thirds its original size, but became flat or inverted shortly after resumption of driving. In S, the lowering of the T-wave in the drive over 300 miles was similar to that noted in shorter drives during the rush hour. (Simonson *et al.* 1968; p. 131)

All the subjects had normal resting and post-exercise ECGs; thus the changes do not seem to be related to oxygen consumption, nor were they related to food intake. Simonson and his colleagues interpret the findings as being due to emotional stress mediated through the autonomic nervous system and hormonal system, particularly the adrenal system. Taggart and Gibbons (1967) in similar vein attribute the S–T- and T-wave changes they found in the ECGs of drivers driving in busy city traffic, to anxiety situations (which they do not specify in any detail). It would be interesting to know what sort of conditions prevailed during the prolonged driving runs; but clearly there is a great need for further experimentation in this area.

5.5. Skin conductance

Palmar skin conductance has been one of the most widely used measures of autonomic arousal. There is an extensive literature on skin conductance recorded during a wide variety of driving situations. This is well reviewed by Helander (1976). Some technical problems are involved in recording skin conductance (to be most accurate it is resistance that is measured; conductance is its reciprocal). The measure is extremely sensitive to movement artefacts, and this is particularly important in towns, for example, where driving controls are more likely to be used. Electrode placement must avoid any interference from steering wheel pressure. Finally, during driving (or any other activity) for any prolonged period, small fluctuations in temperature and humidity may have a considerable effect on basal resistance levels.

There is a well established relationship between skin conductance and traffic events. Michaels (1960) found that 85% of independently counted traffic events generated a measurable GSR; reciprocally, Hulbert (1957) was able to relate 91% of registered GSRs to traffic events. Geometric properties of the road (surface coating, curvature, and objects on the pavement) also account for a large part of the variance in skin conductance (Michaels 1960). Michaels also argues that drivers seem to compensate for low tension-producing environments by driving faster, making their speeds contingent on the perceived complexity of the driving situation.

Taylor (1964) found no relation between skin response and road conditions on a time basis; however, Helander (1976) suggests that this may have been because he averaged his responses over 1000-metre road sections; using such a large unit of distance would tend to obscure variation in the data. When Taylor integrated skin response over distance travelled (rather than per unit of time), he found a matrix of significant correlation coefficients (all between 0·6 and 0·8) which indicated that GSR per mile was highest where accidents were most likely; this also corresponded to where the number of side-turnings per mile was highest and the average speed of the subject was lowest. Taylor suggests that GSR rate acts as a kind of pacing factor: too great a rate of GSRs will signal that a slowing of pace is required. Different drivers will possibly have different pacing characteristics.

Preston (1969), following up this theory, compared the skin response during urban and rural driving, of two groups of drivers with different insurance classifications. The high insurance premium group had a relatively high GSR rate on rural roads compared to the low premium group; on urban roads there was no difference between the groups. Preston makes the suggestion that in towns most GSRs are generated by the actions of other drivers (i.e., it is a paced situation), whereas on an open road they are generated according to a kind of risk-taking level. Brown and Huffman (1972) found that drivers with poor accident and traffic violation records had a higher rate of GSRs than drivers with good records, though this result only had a significance level of $p < 0.25$.

Helander (1976, 1978) reports one of the most detailed and well considered studies of its kind to date; over a relatively small section of road he investigated the relationships between a variety of factors, including geometric properties of the road, traffic events, accident data, vehicle characteristics and movements, drivers' performance and physiological recordings. He concludes that skin response is a very sensitive measure of task complexity and demand on the driver; as he says, "EDR's (electrodermal response) seem to be produced when the relative increase in task demand exceeds a certain value, regardless of the initial level of demand". Babkov (1973) has suggested that there is an optimum level of task demand in a road system that will avoid fatigue due to overloading the driver, or, as he puts it, slowing of the central nervous system.

If skin conductance response can be taken to reflect some kind of pacing factor, either controlling level of risk or level of task demand, what happens to this pacing factor over prolonged periods of driving? Assuming constancy of road and traffic conditions, the general prediction from the hypothesis of progressive de-arousal would be a declining rate and amplitude of GSRs, mediated either by a reduction in speed, or simply by a reduced level of responsiveness to road and traffic events. Attempts to counteract the effect of monotony or drowsiness would result in changes in the opposite direction. Unfortunately, although at least some experimenters have set out to record GSRs (temporary shifts in conductance levels) (e.g., O'Hanlon and Kelley 1974), the majority have reported their results in terms of changes in basal skin conductance levels. One exception is the simulation study of Sussman and Morris (1970). They recorded the frequency of GSR shifts during an emergency-avoidance task, which was incorporated into the simulated driving situation. Although performance accuracy did decline with time, there were no consistent trends for GSR shifts associated with this.

Overall increases in mean basal palmar skin conductance with time were found by O'Hanlon and Kelley (1974) in three out of four experiments investigating devices to prevent running off the road. In the fourth experiment mean conductance level was high to begin with. They are rather puzzled by this result, which seems to run counter to the general trends of their other physiological and behavioural measures. It is possible that these skin response levels reflected directly what was, in some subjects, a marked decline in driving performance. Drifting out of lane across the various lane-marking devices nearly always produced a GSR; this tended to happen in later stages of driving periods (in fact many driving periods had to be terminated due to the danger of the subject falling asleep and leaving the road). Skin conductance could be quite sensitive to this genuine hazard, while other measures would show a low arousal level.

Unfortunately this interpretation would not be supported by results from Dureman and Boden's simulator study. They found a mean increase in skin conductance during the four-hour session for their non-shock group. The shock group, which had a hazard to contend with, showed the opposite trend. A speculative explanation might be found by looking at the subjective rating scales administered: while both groups showed increasing irritation throughout the session (which would be concomitant with increased conductance), only the shock group showed a marked increase in indifference, which might be the basis of the contrary trend, counteracting the increase (see Table 6.4, p. 170).

A more predictable result of declining skin conductance over a four-hour vigilance experiment was obtained by Davies and Krovic (1965). This was associated with a decrease in α activity and increase in θ activity in the EEG, and a decline in auditory vigilance performance. This result is more congruent with the view that skin conductance level is an appropriate measure of tonic arousal. The results from the driving studies suggest that a more differentiated conceptual scheme is needed than the 'arousal' umbrella.

5.6. Electroencephalogram variations

Changes in the electroencephalogram are generally regarded as reflecting the state of attention of the subject. In Shagass's (1972) words: "The most commonly observed EEG correlate of heightened alertness consists of desynchronization of the alpha rhythm with the production of a 'blocking response' of reduced amplitude". Of course, a different EEG response will be produced for different types of activity and different parts of the cortex will produce a different pattern of response. For example, Adey (1969) reports averaged EEG data for 50 subjects resting with eyes closed, and performing both an auditory vigilance task (again with eyes closed), and a visual discrimination task with two levels of time constraint. The EEG signal was subjected to a power/spectral density analysis. At rest, α activity predominated, with a small peak at around 10 Hz recorded from the central and parietal regions and a broader peak from the occipital. Auditory vigilance was associated with a change to a bimodal distribution (10 Hz and 20 Hz peaks) from the tempero-parieto-occipital regions. Visual discrimination produced a marked reduction, from the occipital area, of power in all frequencies above 5 Hz. Power within the α frequency band was also reduced in the central and vertex regions, but in the frontal region there was increased activity throughout the whole spectrum.

Adey claims that it is possible to distinguish between different levels of task difficulty, having found more pronounced changes in the frequency distribution of the EEG in the severe time constraint visual task compared to no time constraint. Creutzfeldt *et al.* (1969) disagree, emphasizing both that there is no simple relation between task performance and the 'blocking' reaction of the occipital α rhythm, and that there is considerable individual variation in response. They found, in studying a variety of mental and visuomotor tasks, that roughly one-third of their subjects exhibited augmentation of occipital α during the tasks, one-third showed a consistent blocking reaction, and the remainder were inconsistent. Mulholland (1969), also critical of the α-blocking attention hypothesis, has proposed the theory that α is inhibited when there is fixation and accommodation of a visual stimulus and that α occurrence should be associated with a decrease of visual accuracy, precision and efficiency: "to the degree that attention is coupled with oculomotor control, to that degree it is liable to be linked with alpha".

Wertheim (1974) has developed a theory that α occurrence is related not so much to visual information-processing, as is implied by Mulholland's theory, but is related to whether oculomotor activity is 'intentive' or 'attentive'. By 'attentive' he means that oculomotor activity is modified by sensory input, whereas during 'intentive' oculomotor activity, internal parameters control it irrespective of concurrent or anticipated sensory information. The theory states that: "the ever ongoing occipital alpha rhythm is attenuated, desynchronized or blocked due to contamination of the EEG trace with electrical activity evoked during activation of the neural mechanism responsible for attentive oculomotor behaviour, while during intentive oculomotor behaviour, the occipital alpha trace remains undisturbed". He relates his theory to prolonged driving: on monotonous stretches of highway very few visual stimuli are relevant to the driving task, and those that are tend to be very predictable; thus attentive oculomotor activity is restricted, implying an increase in α activity. Such a state may be a precondition to falling asleep and may also, he believes, explain the phenomenon of 'highway hypnosis'. In his 1978 report Wertheim gives some experimental evidence in support of his theory. Comparing two tracking tasks, one in which the movement of the target was highly predictable and the other in which it was unpredictable, he found more power in the α band of the EEG in his subjects in the former situation, and this did not seem to be related to the efficiency of visual information-sampling.

Some measure of the amount of α activity, recorded either from the occipital region or between the parietal and occipital regions, has been the basis of the analysis of EEG signals recorded over prolonged driving periods. Various techniques for scoring the amount of α have been used. O'Hanlon and Kelley (1974) and Mackie *et al.* (1974) used four filters to give δ (0·5–2·5 Hz), θ (3–7 Hz), α (8–13 Hz), and β (14–30 Hz) bands. A power-spectral analysis was performed within each frequency band. They also obtained the modal α frequency (as they suggest, this tends to slow as a prelude to sleep). Lecret and Niepold (1974) defined their α index as the percentage increase over resting level of the number of waves within the α band, recorded for five minutes every 15 minutes. Sussman and Morris (1970) used a trained judge to recognize the frequency of α bursts; Kuroki *et al.* (1976) also seem to have used an observational analysis. Sugarman and Cozad (1972) recorded the time spent when the dominant EEG frequency was in the α and θ bands.

Lecret and Pottier's (1971) α index was able to discriminate between a range of driving situations, being highest at night on a closed track, and lowest during daylight city driving. Prolonged night-driving was associated with an increase in the α index; rest pauses had very little effect on this trend. All four of O'Hanlon and Kelley's (1974) experimental groups showed a mean increase in power within the α frequency band, and the majority of the groups showed a slowing of the modal α frequency as the driving runs progressed. One subject who appeared very drowsy, and did in fact fall asleep at the wheel, had more pronounced trends on both of these measures. The researchers from the Calspan Corporation (Sussman and Morris 1970, Sugarman and Cozad 1972) found in their respective studies significant increases in the frequency of α bursts and an increase in the time spent in α associated with time spent at the wheel.

Other studies have not been so successful. Lecret and Niepold (1974) did not find consistent results with the α index; some of their truck drivers showed a decreasing index score throughout the night runs. Studies by Kuroki *et al.* (1976) and Yajima *et al.* (1976) both involved EEG recordings during prolonged driving by day and at night, and while the former study does mention that there were more α waves while driving along straight roads and after lunch, in neither report is there a systematic presentation of results which relates to either time of driving or time of day.

Mackie *et al.* (1974) present their results in terms of power within the combined α and β bands as a percentage of total power. No rationale is given for this, which is strange considering that their co-researchers, O'Hanlon and Kelley, identified increased α with decreased alertness. However, in so far as decreasing percentage α and β power implies increasing percentage power in the θ and lower frequency bands, this index may give a more reliable indication of extreme drowsiness and possibly sleep onset, and with less individual variability than α scores. In one of their experiments, combined α and β power was higher in the second half of the driving session, which runs counter to their hypothesis of decreased arousal; this was the only significant time-on-task trend. They also report that in a hot car and truck cab, percentage α and β power was lower than in a comfortable temperature, and similarly, lower α plus β power was found in a car rather than in a variety of trucks. These results are both interpreted in terms of higher levels of arousal in the cooler temperature and in higher levels of noise and vibration. However, the wisdom of lumping together both α and β bands is questionable considering Sugarman and Cozad's (1972) finding that high acoustic noise when driving was associated with less time spent in both α and θ bands.

Obviously, searching for the elusive optimal EEG measure of the arousal/drowsiness dimension Mackie and Miller (1978) opt for a comparison of power in the θ (5–8 Hz) and β (16–30 Hz) bands in their study of professional truck and bus drivers on irregular schedules. Their main findings can be summarized as follows: many of the changes within trips are congruent with expected circadian variation, thus there was increasing power in the β bandwidth with generally correspondingly less power in the θ bandwidth during many daytime runs; the opposite trend was obtained during many runs that spanned the late night to early morning period.

However, there were important deviations from this pattern. On some night-time trips, particularly on 'sleeper operations' (in which two drivers alternate sleeping and driving en route), there was a pattern of higher β and lower θ power than during other driving periods. There was also a general trend of increasing β and lower θ power as the six-day week of driving progressed in truck relay operations. These trends are interpreted as representing increased cortical arousal mobilized to counteract the fatigue engendered by driving at night, and the cumulative cost of several days of prolonged driving. Interpretation of the trends in EEG power is not helped by the style of presentation and analysis of results in Mackie and Miller's study, which is restricted to comparison of outward and return portions of trips on the same segments of the route, and comparison of (presumably average) levels of the dependent measures in succeeding trips during the week. Contradictory impressions can be gained from these two analyses, particularly in relation to what is happening on the night-shift. It appears that at least some of the fall in 'cortical arousal' that occurs on night-shifts is a decline from initially high levels (that is, high β, low θ power).

However, the data are not presented in a way that permits any thorough analysis or discussion of different influences on EEG trends.

What is interesting about Mackie and Miller's study is the relationship between EEG and other measures of the drivers' state and performance, and the interpretation which is suggested. Fluctuations in the EEG seem to covary rather more closely with steering precision and tracking accuracy, but seem to differ quite markedly at times from fluctuations in subjective ratings of fatigue and performance on a tracking task which was performed before and after driving periods. It is suggested that while the latter reflect a perhaps underlying variation in 'fatigue', the higher 'cortical arousal', particularly during some night-shifts, represents the mobilization of effort which presents a deterioration of actual driving performance (in contrast to performance on a laboratory-style task). It is a pity that Mackie and Miller's data are not more thoroughly analysed to allow such interpretive possibilities to be explored more fully.

Fresh light on the relationship between different EEG rhythms and the driver's psycho-physiological state is provided by some findings from an unpublished study (Thomas and Weiner 1980). In this study, EEG recordings were taken of HGV drivers during day- and night-shifts. The analysis focused on critical episodes during which the driver showed signs of drowsiness, reported either by himself or by an observer in the cab. Such incidents were fairly common on the night-shifts, and occurred in about one in 10 shifts. An EEG analysis was possible in five of these incidents. These varied from a situation in which the driver became so sleepy that he had to stop the vehicle and have a sleep in the bunk, to other situations in which the driver was drowsing on and off on several occasions during a trip and could be seen to be 'nodding off' and then recovering. These critical episodes were characterized by pronounced changes in the power spectral density of the EEG; in the former situation there was a marked peak in the α frequencies; in the latter type of situation peaks in both the α and θ frequencies were typical. These EEG changes are interpreted in terms of a progression towards stage one sleep. Furthermore, the authors suggest that other (more frequent) occasions in which such characteristic EEG changes took place but were not accompanied by overt signs of drowsiness (either reported by the driver or noticed by the observer) represent critical episodes of 'undetected drowsiness'.

While this evidence suggests that increased power in the α and θ bands is associated with drowsiness, confirming and extending O'Hanlon and Kelley's (1974) result, it is not necessarily clear that these EEG changes on their own should be taken as unequivocal evidence for drowsiness.

5.7. *Endocrine response*

There is a large and growing body of literature relating endocrine hormonal response patterns to a range of psychological and social stimuli which concern aspects of stress and coping. There are two, apparently functionally separate, systems involved: the sympathetic adrenomedullary system and the anterior pituitary adrenocortical system. The former, as its name implies, involves reflex sympathetic nervous system control of the release of the catecholamines adrenaline and noradrenaline from the adrenal medulla; this system predominates during the initial emergency response or alarm reaction phase of the general adaptation syndrome (Selye 1956). Excretion of adrenaline in the urine can be taken as a rough quantitative estimate of adreno-medullary activity over a 1–3 hour period. Urinary noradrenaline gives a less reliable estimate of this activity as a large part of the noradrenaline released is reabsorbed by the main nerve endings or bound to various tissues and does not enter the bloodstream or urine. Excretion of both catecholamines has been found to vary with subjective feelings of arousal, both pleasant and unpleasant, and has been shown to be responsive to a variety of stressful situations, including car driving.

The pituitary adrenocortical system is slower acting and less responsive than the sympathetic adrenomedullary system and appears to be particularly involved in the state of resistance to stress of the general adaptation syndrome, though changes in adrenocortical activity do occur in the emergency or alarm phase. The adrenal cortex is under the control of the anterior pituitary gland through the action of ACTH (adrenocorticotropic hormone). In turn it excretes a number of hormones, all of which have been related to the stress response. In the present context the most important of these is the glucocorticoid cortisone which, at stress-related levels raises blood sugar levels, blocks the inflamatory response, interferes with the manufacture of proteins and causes the loss of calcium and phosphate from the kidneys. Brief reviews of these relationships are provided by Frankenhauser (1975, 1981) and Cox (1978).

By far the greatest part of the research on psychosocial influences on endocrine functioning have concerned urinary excretion of catecholamines, and much of the most productive and stimulating of this has been performed by Frankenhauser and her colleagues. Some of their main findings can be summarized as follows.

A range of environmental and task-related factors have been associated with an increased excretion of urinary catecholamines; a typical investigation concerned the effects of monotony and machine control of work in the jobs of sawmill workers. In another study, prolonged working hours in the form of overtime were associated with higher levels of adrenaline excretion after work, during the evening, among a group of female office clerks. Combined with this pattern were feelings of irritability and fatigue, demonstrating the effects of a carry over of work into the leisure hours. Further research suggests that recovery from work is associated with the rapidity of readjustment of the adrenaline response, slow adjustment being associated with poor adaptation to post-work demands and increased fatigue. Other individual differences concerned performance, with high adrenaline (and to a lesser extent noradrenaline) excreters showing better speed, accuracy and endurance on tasks characterized by a low rate of stimulation (such as vigilance tasks), whereas low catecholamine excreters tended to do better on tasks with a high stimulation rate (such as complex choice reaction time). Thus catecholamine excretion appears closely related to the mobilization of effort to cope with task demands; and efficiency of coping and adjustment to various levels of demand and different circumstances may be related to the rapidity with which this adrenomedullary response adapts. However, the research of Frankenhauser and her colleagues also points to a certain specificity of response of different endocrine hormones. Thus noradrenaline has been particularly implicated in physical activity, while adrenaline seems more responsive in situations of emotional involvement or discomfort during heavy physical strain, for example. Also, an interesting dissociation of adrenaline and cortisol response was obtained in an experiment which manipulated control over the pace of work in a laboratory task. Increased control was associated with greater feelings of 'confident task involvement' and a reduction in cortisol excretion, but both adrenaline and noradrenaline excretion increased, reflecting the continued involvement of the subjects in carrying out the task.

Although few of these circumstances are directly comparable to the driving task, the involvement of catecholamine response, in particular, in processes of fatigue, coping and the mobilization of effort, and in reflecting various task demands, should make endocrine response profiles particularly pertinent to studies of prolonged driving. However, research using these variables is only just beginning and results have been mixed.

The most interesting and interpretable results come from Mackie and Miller's (1978) study of truck and bus drivers during regular and irregular schedules, involving both 'sleeper' operations and rotating shifts. Their measures were levels of urinary catecholamines before the start of driving and after each driving period. The most important influence on levels of these hormones appeared to be circadian variation, and trends seemed to be congruent with variation in subjective alertness and performance (particularly on an interpolated tracking task). However, there were occasions when deviations from this pattern occurred. Thus sometimes catecholamine variation appeared to indicate a mobilization of physiological resources to maintain a high level of driving performance; this pattern was characteristic of the group of bus drivers on some trips which ended in the early morning hours. At other times low values of both catecholamines occurred when both performance was poor and ratings of fatigue were highest; this occurred during the final daytime trip of a week of 'sleeper' operations. This perhaps indicates a failure to mobilize sufficient effort to maintain performance, due to the cumulative fatigue induced by six days of alternating sleeping and driving. Other interesting patterns which are worth mentioning were an increasing trend in noradrenaline levels towards the end of the week of driving among regular relay truck drivers, and a tendency for levels of both catecholamines to be higher at the beginning of the working week among relay drivers who were working irregular shifts and who had a loading task to perform. One can interpret the former result as representing a mobilization of effort to counter increasing fatigue, and the latter as a gradual adaptation to the unaccustomed physical work. Surprisingly, in this study, noradrenaline appeared on the whole to be the more labile and responsive of the two catecholamines.

Cullen *et al.* (1979) report no significant variation in catecholamine levels relating to four days of prolonged driving ($11\frac{1}{2}$-hour shifts which started at either 0900 hrs or 1500 hrs). The professional drivers in this study did not, however, show any other evidence of being particularly fatigued or stressed by this driving schedule, which was quite comparable to their normal work. The major exceptions to this pattern were the older drivers in the sample who felt some degree of fatigue during the latter part of the late shift, and they appeared to compensate for this by adopting safer driving margins (a longer time headway to the vehicle in front). Interestingly, one of the main significant findings from the broad endocrine response profile taken from blood and urine samples in this study was an interaction of age and shift with pre- and post-shift cortisol levels. While there was an overall significant decrease in cortisol levels from before to after the shift (which was probably due to circadian variation), this was not so for older drivers on the late shift who had a smaller fall in cortisol than other shift and age group combinations, and which was not statistically significant. This result presumably reflects some physiological cost of continued driving during the early hours of the morning despite their preference to have stopped earlier.

These studies do suggest the value of continuing research with endocrine hormone response measures in elucidating the physiological costs of prolonged driving and irregular shift schedules, but, in themselves, they can only paint a preliminary picture of the relationships which may be found.

5.8. Critical flicker fusion frequencies

Fusion thresholds, both visual and auditory, reflect the state of central nervous system functioning. Systematic variation in these measures has been demonstrated in association both with time of day and time on task. For example, both Grandjean *et al.* (1971) and Kogi and Saito (1971), with air traffic and railway traffic controllers respectively, report a significant circadian variation in critical flicker fusion frequency (CFF) thresholds as well as an effect of time at work superimposed on this. However, the work task and operational stresses of truck drivers are very different from those of traffic controllers.

Results from driving studies have not been consistent. Neither Michaut and Pottier (1964) nor Schiflett *et al.* (1969) found any significant changes in fusion thresholds over three hours' driving. The former recorded visual fusion thresholds before and after road driving; the latter both visual and auditory fusion thresholds before and after simulated driving, with, in one condition, simulated headlight glare. Other studies have reported increases in visual fusion frequencies associated with prolonged driving of up to 10 hours or more, though they have generally not reported the statistical significance of their results (Jones *et al.* 1942, Yajima *et al.* 1976). Ohkubo (1976) reports both increasing and decreasing fusion frequencies for his subjects over a drive of almost 500 km.

5.9. *Somatic measures—electro-oculogram, electromyogram and postural changes*

Electro-oculographic (EOG) techniques have been used both to record rate of eye-blinking and eye movements (horizontal and vertical). Neither Schiflett *et al.* (1969) nor Lecret and Niepold (1974) report any significant variation in eye-blink rate associated with driving time. Lecret and Pottier's (1971) results are more ambiguous: four hours' night driving produced a lower blink rate than two hours', whereas on an all-night drive rest pauses were associated with lower blink rates.

Brown and Huffman (1972) propose that the rate of lateral eye movements (of sufficient angle to include side and rear-view mirrors) provides a good indication of the amount of visual effort involved in driving. Different road and traffic situations reliably produced differences in eye-movement rate. Kuriki *et al.* (1976) also report differences in eye-movement rate for different road types, finding that EOG incidence was lowest on a trunk route after lunch. Increases in EEG α were also found under similar conditions suggesting a post-lunch dip in arousal under visually unstimulating conditions. No other time-related effects are reported.

As a common symptom of drowsiness is head-nodding, a number of researchers have recorded the tension of muscles of the neck which support the head, with the hypothesis that this will decline with the onset of drowsiness. Alerting devices are on the market which are based on the head-nodding principle (see Roberts 1971). Unfortunately the experimental evidence that would favour such a device is not very strong. Lecret and Pottier (1971) report lower electromyogram (EMG) levels in the third and fourth hours of driving than in a two-hour drive, during the night However, three simulation studies during both night and day failed to find any significant variation in EMG levels at all (Schiflett *et al.* 1969, Dureman and Boden 1972, Sussman and Morris 1972).

Also probably related to levels of muscular tension is the hypothesis that prolonged periods in a heavily vibrating truck under conditions associated with tiredness or drowsiness will lead to changes in body posture. Lecret and Niepold (1974) did find that there was an increase in the angle between the driver's back and the back of the seat during prolonged night driving. This suggests an increasingly slumped position of the trunk, which may be associated with muscular fatigue. While a slumped position does attenuate the transmission of vertical vibration through the body, it does make for increased pressure and friction between the vertebrae which is likely to be reflected in the incidence of low back pain.

5.10. *Summary and conclusions*

The majority of studies of drivers' physiological status have been undertaken within the broad framework of the 'arousal hypothesis'. Heart rate has been the measure most frequently recorded and results have been for the most part consistent with the following: short-term changes in heart rate reflect the physical, emotional and attentional demands of the task; long-term changes reflect circadian variation (and perhaps adaptation of circadian rhythms), though there is probably, particularly with some drivers, some effect of time on task, and some effect of adaptation to an arousing experimental situation. Heart-rate variability has not been so sensitive to effects of time; much of the variation in this measure can probably be ascribed to respiration changes. There is considerable scope for further investigation of changes in the S–T- and T-components of the ECG over prolonged driving periods. Skin conductance changes have been found to be very sensitive indicators of minor fluctuations in task demand. Basal skin conductance levels have not shown reliable changes in line with other 'arousal' measures. Both EEG and psychoendocrine hormonal levels have a pronounced circadian variation with increased power in the α and θ bands and lower catecholamine levels during the night. Interestingly, deviations from this pattern, particularly during some late night or early morning shifts or after several days of prolonged driving, suggest an interpretation in terms of mobilization of physiological resources and increased cortical arousal to counteract either the depressed phase of circadian activation or cumulative fatigue. Disturbance of the diurnal cortisol rhythm has also been associated with fatigue. Drowsiness incidents while driving have been associated with transient increases in power in the α and θ bands. CFF thresholds have not indicated consistent changes in cortical arousal. Somatic measures have not produced very dramatic results, though there is potential for development in such measures as EOG and drivers' posture.

Some critical points are pertinent to the general conclusions. Too many studies present only a descriptive analysis of their results without any tests of significance. Often only group means are given without any measure of the dispersion of scores. The insignificance of group results could well be obscuring quite large changes within individuals. Where individual results have been highlighted, it has often been in anecdotal form with no supporting data. Although practical constraints inevitably involved in field experimentation often make a well-balanced experimental design impossible, one suspects that the statistical treatment of results does not in many cases do justice to the results collected.

There are some difficulties in the interpretation of time-related changes in physiological response. Almost all the reported studies of prolonged driving have in their design confounded the effects of time of day and time on task. The studies by Harris *et al.* (1972) and Mackie and Miller (1978) are the major exceptions to this. Secondly, very few studies report any attempt to familiarize their subjects with what is an unusual situation (or certainly should be if the subjects are a naïve random sample). In most cases subjects were used in one experimental session only. Thus, adaptation to the task and to the experimental context cannot be ignored as a possible contributing factor to the results. Similarly, the effects of motivation to do well in this once-off situation may have prevented in many cases a more marked degradation in physiological response. Thirdly, many of the studies had very short experimental sessions—3–4 hours was typical. Significant physiological degradation, particularly in a task like driving, is most likely to be the result of the cumulative effect over several days of prolonged driving, sleep loss, non-driving related daily activities, and, for shiftworkers, the constantly recurring circadian phase shifts (and these are problems that affect not only commercial truck drivers but private and professional car drivers as well). Again, Harris, Mackie and their colleagues provide the most notable exceptions in their studies of sleeper drivers and shiftworkers.

Thus, conclusive interpretation of the changes in physiological parameters over prolonged driving is not possible. It would be fair to most of the studies to say that they were testing the hypothesis that significant degradation in arousal level occurs over prolonged driving periods. In order to do this it is necessary to know what the concept of arousal refers to, and what the criteria are for significant degradation.

Regarding the first of these questions, the evidence against a unitary dimension of arousal has been touched on earlier. Many of the results discussed throughout the chapter confirm the need for a more differentiated theory than the arousal hypothesis. There is evidence of situational stereotypy in heart-rate response (how such a pattern might change over time is not known). Basal skin-conductance levels have shown a marked divergence in direction of change over time from other physiological measures; it may be appropriate to regard this as systematic directional fractionation with skin conductance reflecting perhaps a response to risk or hazard, or emotional factors like irritation or indifference, while heart rate will more accurately reflect general levels of physical and attentional activity. However, there is little evidence with which to fill out this speculation. In a similar way, theoretical work on the interpretation of the EEG, particularly by Wertheim (1974, 1978), points the way to a more developed understanding of physiological changes as function of stimulus parameters, carrying implications for more specific predictions of concomitant behavioural patterns and subjective states.

On the other hand it has been argued that it is the particular concatenation of different physiological parameters that are indicative of the influence of stress or susceptibility to breakdown. There is a large amount of evidence (reviewed by Lavie 1982) supporting the existence of one or more ultradian rhythms of around 90 minutes in duration. An arousal rhythm involves the correlation of a range of physiological (autonomic and central nervous system), sensory and behavioural measures. Lille and Burnod (1982) have shown that this so-called 'basic rest–activity cycle' is susceptible to disruption from occupational constraints. Thus university personnel showed less disruption of this endogenous rhythm than either factory workers, who have to maintain production to a strict time schedule, or air traffic controllers, who suffer transient but often unpredictable demands. Lille (1982) further suggests that the onset of some significant physiological breakdown is characterized by a particular correlation of several physiological measures, whereas variation in one measure may be due to a number of different circumstances.

Thus low catecholamine levels, increasing EEG power in the α and θ bands, a fall in heart rate, and lower responsivity in other measures related to activation such as

GSR and EOG may, in conjunction, be taken to represent declining responsiveness to stimulation from the environment, and a failure to mobilize the physiological resources necessary to maintain alertness and cortical arousal. Sleep or some other form of breakdown or collapse may follow. Unfortunately, few studies of driving fatigue have incorporated a wide range of different physiological measures, and those that have employed a number of measures have not focused their attention sufficiently on critical incidents where the driver is particularly fatigued or drowsy or is performing in some way inadequately. So it is not possible to characterize with any confidence the sort of physiological conjuncture which may be important in terms of driving safety.

Perhaps the clearest situation in which physiological resources are stretched to a limit which may be dangerous occurs when driving is continued despite increasing feelings of fatigue, requiring increasing effort to counteract such feelings. Such a situation might typically occur under a variety of circumstances, including excessive hours of work, lack of sleep and working at night. One would expect feelings of increasing exhaustion to be accompanied by symptoms of the stage of resistance of the body's defence reaction or general adaptation syndrome (Selye 1956); measures of pituitary adrenocortical system functioning might be particularly informative here as an indication of the physiological cost incurred as exhaustion increases. The ultimate point in such a progress would be a state of collapse. However, at what stage and in what ways does this process manifest itself in failure to keep driving safely? For it is clear that drivers can suffer lapses of attention, or other failures of skill or judgement, and occasionally fall asleep at the wheel without reaching a stage that could appropriately be described as exhaustion.

Central to this question is the idea of a dynamic balance between the physiological costs or consequences of work or activity, and the mobilization of resources to compensate for these costs; and crucial to this balance is the individual's motivation to continue activity. Such motivation is embedded in the whole lifestyle and morale of the individual and in his relationship to his work. Thus any physiological account of fatigue and driving safety must on the one hand acknowledge that the driver can to some degree compensate for feelings of fatigue and drowsiness, but that the process of compensation is limited not by some physiological absolute, but by the tolerance that the driver has for the demands of his work over a period of days (or nights), weeks and years, as a matter of routine. The success with which the driver can adapt himself to the demands of his work may be related to the flexibility and predictability of the pattern of work demands and the constraints they place on the driver's own spontaneous rhythms of physiological activation, be they circadian or ultradian. Individual differences may be very important. On the other hand, it must be recognized that the driving task frequently involves contradictory demands on the driver, as when the low level of stimulation which the driver often obtains from his task has to be matched by a high level of attention.

Thus it may not be possible to get any clear single criterion of physiological degradation which might be related to driving safety. However, there are a number of ways in which the relationships between physiological indices and fatigue and driving safety could fruitfully be elucidated further. One involves a fuller understanding of the physiological cost of the job of being a driver, with the shift systems and other demands that it involves, particularly in relation to the disturbance of 'normal' patterns of physiological activity. Secondly, a better understanding of situationally specific patterns of physiological response to particular aspects of the driving task (as, for example, the relationship between EEG and attention has been construed) might enable a more qualitative evaluation of the way the driver is functioning. And thirdly, greater attention to critical incidents of drowsiness and falling asleep at the wheel is needed to establish more clearly the basic physiological parameters of these events and their relationship to other aspects of fatigue and coping with fatigue.

Chapter 6
The Experience of Fatigue

It is an unfortunate fact that comparatively few investigators have taken the trouble to elicit the reactions of their experimental subjects to the supposedly arduous conditions to which they have been exposed, and still fewer investigators have undertaken to explore the experiences of fatigue and drowsiness of drivers in general. This subjective aspect of fatigue is important for several reasons. Firstly, if we are going to ascribe the attribute 'fatigued' to someone, in the final analysis it is he that must decide whether it is an appropriate description or not—it is he that experiences the fatigue, not the experimenter. To rely solely on an operational definition of 'fatigue' (in terms of overt behavioural or physiological phenomena or conditions of exposure) is to change the meaning of the word fatigue as it is used in ordinary language. Secondly, experiences of fatigue are important in their own right; as Cameron (1974) suggests, the most appropriate starting point for research on fatigue is: "a volume of complaint from those who are exposed to a particular set of conditions" (p. 80). Thirdly, in a task like driving, it is the driver's own appraisal of his performance and psychological state which will determine whether he will continue driving, or stop and take a rest when he feels drowsy—it is his reaction to his experiences and feelings that determine his safety. And finally, subjective reports can aid considerably in the interpretation of other measures, behavioural and physiological, as the preceding two chapters, in a rather negative way, amply demonstrate.

Section 2.8 (pp. 56–60) discusses some evidence concerning some professional drivers' experiences of fatigue and drowsiness. This emphasizes most clearly the prevalence of these experiences, particularly drowsiness among shiftworkers. The difficulty of concentrating in monotonous conditions, particularly at night, is very conducive to drowsiness and fatigue, and gives rise to many misperceptions of time and place (like 'missing the miles'), and occasional hallucinations. On the other hand a variety of conditions were considered to be potentially taxing to the driver to a critical extent and these include fog, ice and other bad road and weather conditions. Some drivers stressed that the ability to monitor their subjective state and driving efficiency to enable them to pace their effort and concentration throughout their trip is important in countering fatigue. That people are sensitive to their state of fitness to drive and can make judgements about when they should stop is emphasized by Nelson (1981). However, the problem arises if this judgement is impaired by some other factor such as alcohol or the pressure to complete a journey for professional or business reasons.

The fact that the onset of fatigue and drowsiness is a gradual process with easily recognizable symptoms is emphasized in a much earlier study (Prokop and Prokop 1955). In questionnaires distributed to drivers in motorway cafés, 100 out of 569 people admitted having fallen asleep at the wheel at some time, and 81 of these attributed this to being fatigued. The overwhelming majority of these cases occurred either within two or three hours after midday or in the hours from midnight until about 5 a.m. There was a wide range of factors that were considered to have contributed to fatigue: factors outside the driver and his car included monotonous or well known stretches of road, the weather (rain, heat and sunshine) and time of day (twilight and night-driving); overheating and insufficient ventilation were the most prominent features concerning the car that led to fatigue; and a long list of factors referred to the driver's activity or circumstances (continuous concentration, continuous sitting still, driving alone, previous intellectual work, loading and unloading, physical indisposition, headache, hunger, alcohol, long interruption of journey and slow driving). The first signs of fatigue are also very varied but tend to emphasize the difficulty of maintaining concentration (eye strain and general inattentiveness), the discomfort produced by having to sit in the same position for a long time, a decreasing awareness of the movement of the vehicle and the road environment, a slowing of reaction and a 'roughening' of control movements (Table 6.1 gives a more complete list of the symptoms described). The most effective measures to overcome fatigue apparently involve either stopping (and perhaps sleeping) or taking either a stimulant (coffee) or a sweet drink (Table 6.2).

We shall turn now to consider those few studies that have asked drivers to rate their feelings along certain dimensions during experimental driving sessions.

It is possible that many researchers feel that a subject's self-report statement will be unreliable because that is the nature of subjective processes. To an extent this is an empirical matter and many investigators have found self-ratings to vary in a coherent and systematic way. For example, Dermer and Berscheid (1972) found significant circadian variation in scores on a self-rating scale ranging from extreme alertness, hypersensitivity or excitement to tiredness, boredom and fatigue; this variation very closely matched published accounts of normal deep body temperature variation. Thayer (1967, 1970) also reports significant circadian variation in responses to an adjective check-list tapping again an activation–deactivation dimension. Regarding fatigue due to high physical work load, Hueting and Sarphati (1966) found that

Table 6.1. The first signs of fatigue.

Eyes:	Flickering, pain, feeling of heaviness, vision dimmed and reduced in sharpness, difficulty in fixing gaze, road 'swims', seeing 'objects'
Ears:	Slight drone, hearing ability reduced, sensitivity to noise, humming in the ear
Other feelings:	Feeling of pressure in the head and temples, thirst, hunger, tired back, tired of sitting, painful bottom, general tiredness of body, stiffness, cramp, aches and pains, itching, twitching, sweating hands, lack of air, lassitude
Mental:	Thoughts wander, day-dreaming, concentration goes, inattention, indifference, irritability, impatience, carelessness, weakening of willpower, feeling of intoxication, feeling of well-being, longing for a cigarette, sudden start
Execution of driving:	Delayed reactions, automatic handling of car, increased or decreased speed, awareness of speed lost, traffic signs missed, tired of changing gear, poor gear change, false judgements, driving not straight but zig-zag

After Prokop and Prokop (1955).

Table 6.2. Measures to overcome fatigue.

Measure	No. of replies claiming effectiveness
Stop, interrupt journey	245
Coffee, coca-cola, caffeine tablets, etc.	153
Longer sleep	144
Stop and short sleep ($\frac{1}{4}-\frac{1}{2}$ h)	142
Stop and walk around car	32
Smoking	31
Refreshments	22
Switch on radio, increase volume, change station	14
Eating	11
Conversation with passenger	11
Opening window	9
Singing	7
Cold wash	7

After Prokop and Prokop (1955).

subjective fatigue rated on an unidimensional scale correlated significantly with level of workload.

Several researchers have developed and used fatigue scales in industrial and laboratory situations, for example, Poffenberger (1928), McNelly (1966) and Pearson (1957). The two versions of Pearson's scale are given in Table 6.3. Schiflett et al. (1969) used this scale in their simulation study and found an increase in recorded fatigue between, before and after their simulation sessions. Unfortunately the wider usefulness of this scale is hampered by the cultural nature of many of its items: though possibly in common currency among American Air Force personnel in the 1950s

Table 6.3.   Two forms of Pearson's (1957) Fatigue Scale.

Form A	Form B
1. Like I'm bursting with energy	I never felt fresher
2. Extremely peppy	Extremely lively
3. Very lively	Very fresh
4. Very refreshed	Very rested
5. Quite fresh	Quite fresh
6. Somewhat fresh	Somewhat refreshed
7. Slightly tired	A little tired
8. Slightly pooped	A little pooped
9. Fairly well pooped	Fairly well pooped
10. Petered out	Awfully tired
11. Very tired	Tuckered out
12. Extremely tired	Weary to the bone
13. Ready to drop	Dead tired

from whom it was derived, its carefully constructed equal–interval properties are perhaps lost on the average English-speaker from, for example, the eastern side of the Atlantic.

The Human Factors Research, Inc. group have measured subjective factors in four out of their five major studies of driving. Both Harris *et al.* (1972) and Mackie *et al.* (1974) drew a sharp distinction between variations in alertness and in fatigue. The latter was defined as "a loss of physical strength, a loss of capacity to perform physical work," while the former was related to "boredom, monotony, drowsiness, or simply a lack of attention due to a lack of interest." Both were recorded by means of unidimensional scales; in the 1972 study by 11-point scales, and in the later study by seven-point scales. The scales were administered at the beginning and end of each run, before and after each rest break, and, in the earlier study, after each hour of the run. The procedure in O'Hanlon's (1971) study involved the observer asking the driver at the end of each run what his average level of alertness had been during that run in terms of a seven-point scale with scale point four representing normal alertness.

General trends in both the fatigue and the alertness scale were very similar in Harris *et al.* (1972), with perhaps a slightly more pronounced variation with the fatigue scale. Relay drivers exhibited a slight increase in fatigue over the first six or so hours of driving and alertness remained constant, but both these measures showed an increasing degradation in subjective status (fatigue up, alertness down) after six or seven hours of driving. Sleeper drivers tended to show evidence of more pronounced degradation, starting earlier, than relay drivers, though there was often an 'end-spurt' effect in the opposite direction. Rest breaks usually either temporarily arrested a generally increasing trend, or sometimes effected a slight increase in alertness, but at other times the trip effects were not so consistent. Mackie and Miller (1978), in their follow-up study of irregular schedules, obtained very comparable results with their index of combined fatigue, alertness and sleepiness scales: in all operations there was a tendency for the drivers' subjective state to deteriorate within trips, and for rest breaks to be associated with recovery. Among bus drivers these trends were less marked, but the driving periods were shorter. During rotating shifts and sleeper operations there was a pronounced circadian effect of declining subjective well-being in the late night/early morning trips. Having to perform a moderately heavy loading task seemed to have no effect on this measure.

In general terms these findings give support to the hypotheses of Harris, Mackie and their colleagues, and show similar directional trends to many of the physiological recordings (in particular heart rate, less so the EEG—see Section 5.6). There were some other discrepancies, for example declining heart rate during the first seven hours of relay driving was not matched by any change in alertness ratings (suggesting adaptation rather than any 'fall-off' in arousal as the more appropriate interpretation). Unfortunately there was no inferential statistical treatment of results and no exploration of individual variation of subjective ratings and physiological response, so a detailed evaluation is not possible. There were some drivers who were judged by the observers to be very tired or fatigued; these tended to have larger than average overall decreases in heart rate with very little recovery from rest breaks. It would have been interesting to know whether subjective self-ratings reflected these differences.

In Mackie *et al.*'s (1974) study, trends for fatigue and alertness were again similar, both showing significant degradation in subjective status over time, in most but not all conditions. The mean effect was greater in a high temperature rather than in comfort, though this difference was not statistically significant. Overall differences between noise and vibration conditions were not significant, though with drivers acting as their own experimental control between truck and car driving, truck driving produced higher fatigue ratings and lower alertness ratings than car driving; and the latter finding, in particular, seemed to run counter to the authors' interpretation of physiological records and the drivers' expressed feelings of boredom. O'Hanlon (1971) gives very little detail of the results of his drivers' self-ratings except to note that there was a general lack of correspondence between rated alertness and either heart-rate variability or lane-drift frequency; he gives as an example of this an occasion when, just before a subject dropped off to sleep, he rated himself as normally alert, though both HRV and lane-drift frequency were abnormally high.

Attempting to differentiate further than the Human Factors Research dichotomous construct, Dureman and Boden (1972) required their subjects to rate the felt intensity of eight subjective variables (see Table 6.4), rating from 0 to 100, with 10 representing just a slight feeling, and 100 'the greatest intensity I ever felt'. Rated intensity of all these variables showed a marked increase throughout the simulated driving session. This was particularly so for boredom, though irritation, 'cautiousless-ness', sleepiness and drowsiness also showed large increases. Those in the shock group, not surprisingly, did not show a high degree of sleepiness.

Perhaps the most comprehensive set of subjective rating scales to have been used in a study of 'fatigue' and driving was used by Fuller (1978) and McDonald (1978) in an experimental investigation of drivers' following performance during long periods of

Table 6.4.   *Medians of subjective ratings at the end of the first and the fourth hour of a simulated driving task. Ratings were taken hourly.*

Rating	Non-shock group		Shock group[a]	
	1st rating	4th rating	1st rating	4th rating
1. Bored	17·5	62·5	22·5	65·0
2. Irritated	2·5	42·5	5·0	50·0
3. Cautiousless	20·0	57·5	25·0	55·0
4. Sleepy	45·0	77·5	20·0	50·0
5. Drowsy	35·0	80·0	32·5	30·0
6. A feeling of unreality	27·5	32·5	30·0	17·5
7. Difficulty in keeping eyes open	42·5	57·5	32·5	40·0
8. Indifference	45·0	50·0	25·0	57·0

$N = 8$.
[a] In the shock group small electric shocks were administered to the subjects contingent upon driving errors.
After Dureman and Boden (1972).

driving. They asked the drivers to rate their performance and feelings retrospectively, hour by hour, after each half of an 11-hour driving shift. The rating dimensions are listed in Table 6.5; they include various aspects of driving performance as well as a number of emotional and feeling-tone dimensions. The scales were based on an analysis of driving into what seemed to be its most important component skills, and on an interpretation of previous studies of drivers' experiences of fatigue and drowsiness at the wheel. Drivers were also asked to describe the nature of any adverse feelings and to rate their fitness to continue driving at the end of each driving session.

Table 6.5.   *Rating dimensions of Fuller (1978).*

Driving ability
Observation
Control of vehicle
Decision making
Courtesy to other road users
Riskiness
Drowsiness
Boredom
How quickly time seemed to pass
Irritation
Exhaustion
Physical discomfort
Unaware of what you were doing
Unaware of time passing
Day-dreaming
Seeing things which were not there

Unfortunately the richness and detail of the design were not matched by richness and sensitivity of results. There was very little variation in the drivers' ratings, and it was clear that the experimental requirements were well within the untroubled capability of the drivers, for whom these conditions were fairly typical of their daily routine. There were some slight but significant differences between different groups of drivers, however: thus, for example, drivers on the late shift (1500–0230 h) were more drowsy, bored and uncomfortable than drivers on the early shift (0900–2030 h); they also tended to rate their general driving ability and control of the vehicle as worse and be less prepared to continue driving at the end of the shift. Older drivers tended to rate their general driving ability as worse and more risky than younger drivers did, and to suffer more drowsiness and irritation. A subsequent study in which the drivers were required to follow another vehicle for the entire driving shift showed that this change in conditions was associated with slightly higher levels of 'fatigue' (in terms of drowsiness, discomfort and the desire to have stopped earlier, for example). Again, older drivers were more affected by drowsiness on the late shift (Fuller 1983).

These studies represent an attempt to provide a more detailed and sensitive account of experiences of fatigue than had previously been undertaken. The fact that so little variation was displayed suggests that a qualitative evaluation of circumstances where the driver feels he is functioning less than adequately may be a more useful strategy than the quantitative monitoring of repeated days of more or less normal uneventful activity.

All these multi-dimensional scales are based on intuitive conceptual distinctions, albeit appearing very logical. It has also proved possible to empirically identify several independent factors within the fatigue complex. Wolf (1967) identified three major factors from an analysis of essays on fatigue. These were nervous fatigue, drowsy fatigue and physical exhaustion. He hypothesized that these would be related both to type of task and level of motivation, according to the scheme in Table 6.6. He was able to show that the three fatigue scales he constructed were discriminatively different, but they were not related to different types of task; however they were related to motivation level. Saito *et al.* (1970) similarly found three major factors which formed the basis of the fatigue scale of the Industrial Fatigue Research Committee of Japan. The items on the checklist are given in Table 6.7. The first factor relates to drowsiness

Table 6.6. Hypothesized relationship between type of fatigue and motivation and type of task.

Motivation	Task	
	Sedentary	Non-sedentary
Low	Drowsy	
High	Nervous	Exhaustion

After Wolf (1967).

Table 6.7. Fatigue check-list.

(A) Drowsiness and dullness	(B) Decline of working motivation	(C) Projection of fatigue to some part of the body
1. Feel heavy in the head	11. Find difficulty in thinking	21. Have a headache
2. Feel tired in the whole body	12. Become weary while talking	22. Feel stiff in the shoulders
3. Feel tired in the legs	13. Become nervous	23. Feel a pain in the waist
4. Give a yawn	14. Unable to concentrate attention	24. Feel constrained in breathing
5. Feel the brain hot or muddled	15. Unable to have an interest in thinking	25. Feel thirsty
6. Become drowsy	16. Become apt to forget things	26. Have a husky voice
7. Feel strained in the eyes	17. Lack of self-confidence	27. Have dizziness
8. Become rigid or clumsy in motion	18. Anxious about things	28. Have a spasm of eyelids
9. Feel unsteady while standing	19. Unable to straighten up in posture	29. Have a tremor in the limbs
10. Want to lie down	20. Lack patience	30. Feel ill

After Saito *et al.* (1970).

and dullness, the second particularly to motivational aspects, attention and concentration, and the third to physical symptoms.

Yoshitake (1971) explored the interesting idea of relating the multi-dimensional symptomatology of fatigue to its intensity of feeling, which he conceptualized as a single dimension characterized by 'overall unpleasantness'. He compared responses to the checklist with scores on a nine-point scale ranging from 'feeling fit, rested' to 'feeling extremely tired, exhausted'. He found a very strong linear relationship between the number of symptoms felt and the score on the rating scale. This was true up to about seven or eight on the rating scale; with higher scores the number of complaints increased steeply. In terms of the weighted contribution to the overall feeling of fatigue of the different types of symptoms, somatic symptoms (C) (see Table 6.7) had the greatest effect, though the number of drowsiness symptoms (A) was the greatest; motivational symptoms (B) occurred without some physical symptoms. This general result does suggest that one can estimate the degree of fatigue by counting the number of symptoms.

Yoshitake's subjects were bank workers and broadcasting workers. Different patterns of response and possibly different weightings might be expected from a group like lorry drivers. One might expect that particular problems like back pain and eye strain would be important. A rigorous qualitative examination of the subjective state of a group like professional truck drivers would potentially add considerably more to understanding their operating problems than a numerical score corresponding to a point on a general fatigue scale. When one compares the fatigue and alertness rating scales of the HFR group with the dimensions of the Japanese

Fatigue Scale, the neatest fit seems to be to identify 'alertness' with a combined A and B dimension and 'fatigue' with the C factor. However, it is possible that this might be doing violence to the conceptual scheme of Harris, Mackie and their associates, and it is certainly not clear that this is how their truck drivers would have interpreted the scales.

It is also possible that observers can be sensitive recorders of an operator's subjective state. After all, we are all well practised at intuitive judgements of others' psychological state. Kashiwagi (1971) has developed a Fatigue Rating Scale; this scale has two dimensions (derived by factor analysis)—weakened activation and weakened motivation—which can be identified fairly closely with the first two factors on the original fatigue checklist discussed above. Table 6.8 shows the items of the scale together with their factor loadings. Kashiwagi states: "One seems to be able to discriminate between the dull appearance with a sulky listless face and the sluggish or restless appearance with a pale stiff face—"does one? The scale when applied to railway engine drivers showed an increase in mean number of symptoms after work compared to before.

It should be pointed out that few of the times are at all precise in physical or behavioural terms and many of them seem to require a fairly imaginative interpretation (e.g., 'spiritless eyes'). Rating scales such as this are an important development and particularly useful when requiring the operator to fill in a self-rating scale will interfere with his task, or possibly sensitize him unduly to fatigue effects. However it is particularly important to know the reliability of a scale like this between administrations by different raters, as well as its validity in terms of the criterion of the examinees own self-ratings.

The problem with nearly all the rating scales that have been used with drivers in these experiments is that they are extremely difficult to interpret. If a driver moves his rating down one or two points what does this actually mean? Very little. The studies

Table 6.8. *Items for the Fatigue Rating Scale of Kashiwagi (1971).*

Items indicating 'weakened activation'	Factor loading	Items indicating 'weakened motivation'	Factor loading
1. Too lazy to walk	0·857	1. Many mis-statements	0·916
2. Unsteady voice[a]	0·754	2. Avoid others' eye	0·910
3. Absent minded	0·700	3. Difficult to speak to	0·886
4. Hollow cheeked	0·669	4. Sluggish	0·850
5. Avoid conversation[a]	0·571	5. Restless	0·784
6. Sulky face[a]	0·568	6. Anxious about other things	0·783
7. Spiritless eyes	0·520	7. Pale face	0·736
8. Irritable	0·514	8. Stiff face	0·733
9. Listless face	0·510	9. Trembling fingers	0·718
10. Dull	0·500	10. Unable to concentrate and listen	0·701

[a] These items also have high factor loading on the second component.

discussed at the beginning of this chapter emphasize that the causes and symptoms of fatigue are many and various, but there does seem to be a close relationship between the different types of contributory factors and the symptoms—not surprisingly (the problem of concentration, for example). These studies also suggested the importance of the driver recognizing these symptoms and acting upon them. The complexities of this process that the driver performs in monitoring his experiences and feelings is in no way adequately represented by him placing a tick somewhere between 1 and 10 on a continuum of 'fatigue' or 'arousal'. It is only through asking the driver for a more qualitative and descriptive account of what he is thinking or feeling that it will be possible to pinpoint what aspects of his task, his job in general, or his lifestyle, are causing him stress, in what ways is the quality of his driving likely to deteriorate, or his health suffer, and what might be the most appropriate countermeasures. The interview and questionnaire studies discussed at the beginning of this chapter have not the range and scope to give full and adequate treatment to driving fatigue, particularly as it affects the truck driver. The experimental studies employing ratings scales have not appreciably advanced our understanding of the dynamics of the experience of fatigue, though the exploration of the various factors that comprise fatigue represent an improvement in this type of study.

The lack of a systematic treatment of the experience of fatigue on any extensive scale betrays a fairly pervasive prejudice concerning the scientific status of subjective reports. This is based on the ideas that (*a*) there is no independent check upon the accuracy of subjective reports, (*b*) the act of reporting may generate artefactual memories etc., and (*c*) such reports are too vague to be of much use. In contrast to this it is clear that subjective statements can indicate, clearly and accurately, states of mind; it is impossible to give an account of someone's state of readiness for activity without taking notice of such states of mind; the distinction between statements about observable behaviour and unobservable feelings (upon which this prejudice is based) is by no means a sharp and distinct one (Harré and Secord, 1972, give a more detailed and comprehensive discussion of these points). If you want to know about fatigue and driving it seems sensible to ask a driver: too few investigators have taken such a seemingly elementary step.

Chapter 7
Conclusions and Criticisms

7.1. Introduction

This final chapter begins with a brief résumé of the main conclusions that can be drawn from the evidence discussed in the earlier chapters. This is followed by a few critical points of a general nature, an outline of the development of research into 'fatigue' and driving, and an outline of some of the practical implications of this research.

7.2. Main conclusions

Very long hours of work and prolonged periods of driving characterize the working conditions of a large number of truck drivers. For many drivers (particularly those in long-distance or international haulage) the normal working week is comprised entirely of time devoted to work and time for recuperation from work (food, sleep, hygiene) and in many cases time for the latter must be severely curtailed. Domestic and social activities are restricted to the weekend. Preserving at least this minimum time for social and family life seems to be at least one factor which makes for pressures to finish a particular driving assignment even though this may mean exceeding both the legal restrictions on driving time and limitations on the driver's own ability to drive safely and well. Shiftworking is widespread within the haulage industry; a variety of shift systems are operated including permanent night-shifts, alternation between days and nights, various forms of rotation, and 'sleeper' operations) where two drivers alternate driving and sleep around the clock). Irregular shift schedules inevitably involve disturbance of sleep patterns, a curtailment of sleep, and driving during periods of the day when circadian physiological activation is more appropriate to sleep than work. In such circumstances the driver may be subject to the combined effects of an accumulated sleep deficit, prolonged driving and other work activities, and driving during the depressed phase of the circadian rhythm. The driver's difficulties in obtaining sleep are compounded by having frequently, and regularly, to sleep in the truck cab, which affords an inadequate environment to obtain satisfactory sleep, being subject to disturbances of temperature, noise, light and vibration.

Drivers themselves do experience these conditions as stressful and fatiguing. Prolonged hours of driving are associated with a higher level of fatigue, and among shiftworkers drowsiness while driving is a very prevalent complaint. Many drivers also experience drowsiness in the post-lunch/early afternoon period. Fatigue and drowsiness however are not seen by drivers as only reflecting the effects of time of day and time at work, but as resulting from a range of environmental and job-related factors, including, for example, bad weather, hot weather, heavy traffic or monotonous motorway driving. The stresses of work do seem to structure, at least in part, the development of the truck driver's career, with a tendency to gravitate towards easier, more regular and less variable work. Drivers, almost universally, value their work for its relative autonomy and freedom from managerial control.

Truck driving is a dangerous occupation in relation to other occupations. Most of this danger stems from the threat of death or serious injury from traffic accidents (as well as other accidents, which are presumably related to the driver's duties of loading or otherwise supervising his vehicle). There is also a higher than expected occupationally related incidence of death from lung cancer, which may be due to an association between smoking and the job of driving. The driver's health is also at risk in some respects from other hazards involved in work. Thus, drivers are as prone as other workers to disorders of adaptation to shiftwork, which tend to be most severe and common in their effects on sleep and digestive processes as well as affecting mood. In the wider sphere of psychosocial health, shiftworking can also be detrimental to satisfactory domestic and social functioning. The physical environment of the truck cab also contains potentially damaging hazards, the most serious of which are noise and whole-body vibration. Levels of noise typical of many truck cabs under normal operating conditions exceed 90 dB(A) in many cases, and 85 dB(A) and over is very common. A lifetime of exposure to such levels does pose a threat to the truck driver's hearing and some drivers do suffer from a significant hearing impairment. The use of personal hearing protectors is inappropriate in truck driving under normal circum- stances. Levels of whole-body vibration in truck cabs tend to be highest near the frequencies of maximal human sensitivity. Levels are variable (depending on the road, and chassis, cab or seat suspension) and under typical US conditions may vary between within the four-hour and within the one-hour Fatigue Decreased Profici- ency boundaries of the ISO limits. When combined with the strain of maintaining a constant .posture (particularly with inadequate seating support) and the muscular strain involved in driving, these levels of vibration may contribute to back pain and the exacerbation of lumbar spinal disorders. It is not clear whether these stresses are sufficient in themselves to cause spinal damage or whether this may arise from shock or impact from major road irregularities. Concerning both noise and vibration, it is clear that engineering solutions to vastly reduce the level of these stresses have been available for some time before their application becoming the norm in the vehicle manufacturing industry. While it seems that increasing attention is being paid to the truck cab environment it may be some years before existing ergonomically inadequate cabs become obsolete.

There is some evidence that truck and bus drivers may suffer higher rates of a range of digestive, circulatory and musculoskeletal disorders in comparison to other occupations. In relation to truck drivers nervous stomach, haemorrhoids, healed peptic ulcers and appendicitis have been implicated. It is plausible to suggest that a range of stresses characteristic of the working conditions of truck drivers may be involved in the aetiology of these disorders, though this can only be considered tentative and speculative. Such stresses would include vibration, the need to maintain a constant posture, lack of physical exercise, the lack of an adequate diet, the disruption of diet and digestion by irregular work schedules, and the mental stress involved in the job of driving.

It is not easy to get an accurate picture of the extent to which conditions associated with 'fatigue' make a contribution to accidents involving heavy goods vehicles. Estimates of the proportion of accidents where such conditions have made an identifiable contribution vary from 1% to 7%. Such estimates cannot be taken as a reliable guide to the extent of the contribution of fatigue; they do not take into account the hidden but probably pervasive influence of fatigue on many accidents which result, at least partially, from slight inadequacies of judgement, visual search or anticipation, where the scope for error is enormous and human abilities do not match very well the demands of the driving task. There is a marked circadian periodicity in 'falling asleep at the wheel' accidents with a disproportionate number occurring during the night-time hours. They do not necessarily occur after prolonged driving. There is a similar circadian distribution of single-vehicle accidents and many of these involve dozing at the wheel. Drivers of trucks are most at risk of death or serious injury in single-vehicle accidents, while in collisions involving other vehicles it is the occupants of the other vehicle who are most at risk.

The accident involvement rate of trucks (in relation to the amount of truck traffic) has been found to be around twice as high in the night-time hours (2000 hrs to 0600 hrs) than during the daytime hours, and $2\frac{1}{2}$ to 3 times higher after 14 hours of work than after less than 10 hours of work. There is a strong association between hours of driving and accident rates, with about twice as many accidents occurring in the second half as in the first half of trips. However, there is no clear pattern establishing a rising accident rate after so many hours of driving. Part of the problem may be the inevitable confounding of factors to do with hours of work or driving with those to do with time of day, with the added complication of the timing and duration of sleep. Thus different relationships between hours of driving and accident rates may obtain for different shifts. One potentially dangerous effect of shift on accident frequency which has not received attention in the truck accident studies is the influence of early morning shift starts on accident frequencies during the day, which has been established in one study of bus accidents.

One of the clearest conclusions to emerge from the experimental studies of driving performance and physiological status is that there is no inevitable deterioration in driving performance or physiological state which results from prolonged driving *per se*. Numerous studies have demonstrated that it is possible for many drivers to continue driving or simulated driving with no identifiable change in performance which is interpretable as a deterioration, and with a normal variation of at least some physiological measures. Nevertheless, there are circumstances where deteriorations in driving performance and physiological changes indicative of very low arousal states have frequently and fairly reliably occurred. Such circumstances include, in particular, after the accumulation of several days of prolonged driving, during driving at night, and following some degree of sleep deprivation. This is not to say that performance decrements simply do not occur after long hours of driving, for they have, on occasions, been found; rather, these decrements are more likely under the above mentioned circumstances.

There are a variety of ways in which driving performance has been found to deteriorate under conditions associated with 'fatigue'. In many of these there is some corroboration between drivers' own reports of the effects of fatigue and experimental evidence measuring aspects of performance. Thus drivers' complaints of a loss of a sense of speed, or of misperceptions of velocity are congruent with changes in the patterning of eye movements. Similarly, the experience of loss of concentration and alertness is matched by experimental findings of reduced vigilance and the missing of signals and road-signs. Again both sources of evidence have pointed to increasing reaction times and a coarsening of control movements which are related to a decline a tracking accuracy and an increasing tendency to 'lane wander'. Feelings of impatience, carelessness and distorted judgement are mirrored by some findings of riskier decisions and less courtesy to other road users. Some reports of very close following by fatigued drivers have not been reproduced experimentally. There are also reports of misperceptions of time and place, visual distortions and hallucinations. The situations under which these phenomena are likely to occur might also be characterized by physiological changes typifying low levels of arousal (particularly low heart-rate levels) and changes in the EEG which increasingly approximate to stage 1 sleep.

There are a number of circumstances which appear to exacerbate feelings of fatigue or drowsiness, or which may otherwise make for a deterioration in the truck driver's performance. These include long monotonous stretches of road, heavy traffic congestion and delays, bad weather, hot weather or cab temperature, and having eaten a large meal. Both noise and vibration experienced at high levels over prolonged periods, and extremes of temperature, are likely to contribute to feelings of discomfort, fatigue, annoyance and irritability, though such effects are better documented in relation to noise than vibration or temperature. High levels of noise will mask auditory signals emanating from other traffic or the truck itself. Noise may also contribute to a loss of attention, particularly in paced aspects of the driving task and to items in the periphery rather than centrally in the driver's field of attention. Perceptual judgement may become more variable and include more errors. However, these findings are not based on studies of driving. Whole-body vibration may affect visual acuity to a small extent (use of mirrors may be most affected) and steering accuracy (also to a small extent).

7.3. General criticisms

It must be acknowledged that there are discrepancies and inconsistencies within these findings and it would be wrong to imply a high degree of uniformity in the evidence received. Many of these stem from inadequacies in the design of and analysis of investigations. Hence it is appropriate to make some general critical comments on the research.

There has been too much of an hiatus between controlled experimental studies and the research that investigates and outlines the actual conditions under which driving is carried out in 'real life'. Admittedly the bulk of experimental research on 'fatigue' and driving has not been particularly concerned with the commercial truck driver, nevertheless it has also exhibited a lack of interest in what non-professional drivers do outside the laboratory. Correspondingly, the social and normative research describing the nature and extent of the 'fatigue' problem has tended to be rudimentary and fragmented. A major exception has been the series of major studies on truck and bus driver safety and performance by Mackie, Harris and their associates of Human Factors Research, Inc.

Little consideration has been given to the social dynamics of the experimental situation; this must have a crucial bearing on a problem like 'fatigue' which is very much influenced by social norms and expectations, by motivation and attitude to the task, as well as by the more easily definable (and measurable) physical and temporal parameters of the situation. Experimental results have tended to be presented averaged over a group of subjects, and this is bound to have obscured the extent of individual variation in susceptibility to the stresses of the experiment. The effects of such stresses must be mediated by so many factors specific to each subject that it is difficult to imagine that averaged group results could be very meaningful, or definitive of the state of fatigue.

The bulk of the experimental literature has ignored the complexity of the stresses, social and physical, to which the driver, professional or not, is potentially exposed. Time driving, with or without rest breaks, has been the most common independent variable, with the occasional inclusion of time of day or sleep deprivation. Thus, to a large extent, these studies describe a fairly normal pattern of activity, where the subject is driving, albeit for long periods, well within his capabilities, particularly if it is a one-off occasion that he does not have to repeat virtually every day. Little attention has been paid to those situations in which the driver is struggling to keep himself going, perhaps afflicted by an accumulated sleep deficit, driving in bad weather, with a driving schedule that prevents him taking adequate rest; and where he may be oppressed by domestic troubles or a dispute with his employer, or any other combination of a host of circumstances that can turn a relatively 'normal' 10-hour driving day into a constant struggle to keep concentrating, to stay awake, and generally to perform adequately behind the wheel.

Finally, few of the experimental studies have considered their dependent variables in terms of safety. This is true of the majority of studies which have recorded some aspect of driving skill, as much as of those which have taken physiological recordings or asked their subjects to fill in rating scales describing their feelings. Thus any attempt to relate these studies to the accident causation process must involve considerable conceptual leaps.

7.4. A brief history of fatigue research

Many of the inadequacies of research into 'fatigue' and driving can be put into a slightly more comprehensive context by a brief consideration of its history over the past 40 years or so. In very general terms, three phases of research can be distinguished according to fairly marked differences in general approach to the problem.

Typical of the earlier period of research are the studies of Ryan and Warner (1936), Lahy (1937) and Jones et al. (1941). Their characteristic method of studying 'fatigue' in the driver was to require him to perform a range of psychological and physiological tests after certain periods of driving. These included various types of reaction–time test, tests of hand steadiness, body sway, blood pressure, etc. Driving performance itself was not measured, but these tests purported to measure abilities or attributes crucial to effective driving. Johnson's (1945) critique of the Jones et al. study applies equally well in most respects to the others and is very damaging to their claims to validity. First of all, Jones et al.'s study is taken to task for failing to attempt to establish any relationship between accidents and hours of driving; next it is criticized for failing to define what the investigators mean by 'fatigue' independently of the various tests that supposedly measured it; the adequacy of the research design is questioned; the relevance of the tests used to the task of driving is doubted; and the possibility of relating the statistically derived index of fatigue in any meaningful way to driving safety is denied.

It is hard to know how influential Johnson's paper was, but work in the next identifiable phase of research, which flowered from around the mid-1960s, showed an improvement in one respect—the performance measures used appeared to be rather more relevant to the activity of driving than those in previous studies. Important in this period of research was Platt's (1964) theoretical paper, and the associated work which was largely directed towards the automatic recording of drivers' control movements. Also around this time there were a number of studies of vigilance and driving, importing into the car or simulator the apparatus and methodology of signal detection experiments. In Britain, I. D. Brown was conducting his research with subsidiary tasks into the 'spare mental capacity' of drivers. Although these studies had a certain degree of face validity, the major research effort was still directed towards finding a reliable index of 'fatigue', and one of the major criticisms, articulated by A. D. Little, Inc. (1966), of these studies was their almost total lack of concern with accident causation. McFarland and Mosely's (1954) report on near-accidents was the only attempt by this time to relate hours of driving to safety. The arousal hypothesis was very popular and correlations between declining measures of arousal and changes in performance could, it was supposed, be extrapolated to extreme 'de-arousal' (whatever that may be), sleep and/or disintegration of driving skill. Again, although some reviews (Crawford 1961, Brown 1967) emphasized that the causes of 'fatigue' included far more than just hours of driving, few investigators seemed to pay much heed, or show much interest in whether the experimental conditions they had set up were typical of those affecting significant numbers of drivers outside the laboratory.

It is possible to identify the beginnings of a third phase of fatigue research in work published since the early 1970s. This work is characterized by an increasing diversity in topic, methodology and theory, a recognition of the failings of previous research to establish any substantial findings, and the realization that the roots of this failure lay in the inadequacy of the theoretical basis of fatigue research. Cameron, in his 1973 and 1974 analyses, emphasized that fatigue cannot be taken as some kind of hypothetical intervening variable with causal status, that it is not so much hours of driving but the whole life pattern of the individual that underlies the 'fatigue' problem, and that 'fatigue' is no more than a convenient label for a generalized response to exposure to stress over a period of time, and that this stress response is not specific to 'fatigue'. This work can be seen as representing the beginning of a radical reappraisal of fatigue research.

Perhaps occupying a transitional position between these two later phases is the series of research reports produced by Human Factors Research, Inc of California (see Harris *et al.* 1972, Mackie *et al.* 1974, O'Hanlon and Kelly 1974, Mackie and Miller 1978). The broad scope of the research in these reports is unprecedented, encompassing studied of accident statistics, surveys of working hours and of drivers' attitudes to their conditions of work, field studies in instrumented vehicles investigating a range of commercial operating conditions and various types and levels of environmental stressors, and the evaluation of possible countermeasures to running off the road. However, the experimental approach adopted in this series of studies is very much characteristic of the second phase of research discussed above. The dependent measures are treated very much as 'measures of fatigue' with little analysis of what they actually mean; whether they be steering wheel reversal rate, physiological measures, or scales of alertness. Their results are also analysed as group results with little emphasis placed on individual variation. It is as if fatigue was hypothesized as some kind of mediating variable, which is indirectly measured by changes in performance, physiology or subjective state; and which has a direct and linear relationship with the major independent variables like hours of driving and time of day.

Admittedly these studies were designed primarily to produce fairly definite answers to pragmatically phrased questions about the adequacy of legislation in preventing fatigue (among other things); hence the focus is not on a theoretical elaboration of the nature of fatigue. However, it does mean that there are major conceptual problems in actually testing an hypothesis like: "that significant fatigue may occur within the limits of current DOT regulations". There is little discussion of what might be defining criteria of a 'state of fatigue', nor of how to interpret the dependent measures in terms of safety which is surely the crucial issue. One cannot help feeling that there was more scope in these studies than was given expression for the analysis of situations where a driver may have appeared to be more critically impaired physiologically or at risk in terms of performance adequacy. There could also have been a more qualitative analysis of the driver's subjective state on such occasions which might have given a more meaningful idea of how he felt, and a more detailed quantitative analysis of the relationships between different dependent measures, which might have given some idea of the extent to which they were measuring aspects of the same thing.

The possibility of a theoretical orientation of research into fatigue and driving depends on a number of different avenues of research. Firstly, it is necessary to have a thorough understanding of what working conditions actually obtain within the road haulage industry and how drivers themselves understand and respond to these conditions. In the non-professional sphere one would want to know about the ways in which people plan and organize their driving and the constraints under which they operate.

Secondly, it is necessary to establish as clear a connection as possible between these conditions and actual accident risk. Then, to understand the process which intervenes between the conditions of life and work and accident levels, it is necessary to try to develop measures of driving performance and physiological state which are in fact related to critical incidents of performance failure like falling asleep, failing to notice another vehicle on the road, or making faulty decision in some manoeuvre. Also one needs to be able to conceptualize the relationships between different types of measure, as, for example, the relationship between EEG and visual attention has been hypothesized.

However, central to an understanding of the process of fatigue is the driver's own experience of fatigue, because what happens depends on the driver's actions (in relation to the road and traffic situation) and such actions can only be understood in terms of the driver's awareness of the situation, even if they cannot be totally explained in terms of that awareness.

It is clear that a substantial beginning has been made in answering a lot of these questions, but a satisfactory theoretical account of fatigue and driving is a long way off. However, there are a variety of issues which crucially affect the interpretation of research results which deserve attention.

One issue is that of whether the driver can, or does, compensate for feelings of fatigue, or physiological decline, and maintain performance at a safe and skilful level, and what the limits of that compensation are. Thus studies of following performance have indicated an increase in safety margins associated with feelings of fatigue (which were not particularly pronounced). Similarly, EEG changes indicative of cortical arousal have been associated with times when driving performance has been maintained at a high level despite declines in both subjective state and performance on an interpolated laboratory-style task, suggesting perhaps some mobilization of effort to overcome fatigue. However, there are presumably limits to this compensatory effort and a particularly profound decline in performance and physiological and subjective state may follow if driving is continued beyond the effective limit of compensation. How suddenly such a breakdown in compensation may occur is a matter for speculation, and may depend on circumstances. The experience of drowsiness at the wheel seems sometimes to be a gradual process; on other occasions, it seems, sleep can take over relatively rapidly and unpredictably.

In all of this, changes in motivation concomitant with fatigue may be of crucial importance—thus at some stages increasing safety margins seem to serve the function of avoiding the mental effort of risky decisions or difficult perceptual judgements; on others the desire to stop or to finish the driving assignment by getting to the driving destination as fast as possible may distort judgemental process, by, for example, changing decision criteria, or changing the evaluation or acceptance of risk; and, ultimately, the interpretation of hypnagogic hallucinations as expressions of a strongly fought against wish-fulfilment to stop also points to the importance of motivation in these potentially fatal circumstances. One important implication of the driver's ability to compensate for fatigue is that even though performance may be maintained at a high level there will nevertheless be some physiological or psychological cost to the driver which will persist after the driving period and may carry forward to other driving periods. This may have implications for sleep and recuperation from work; it certainly implies the possibility of cumulative fatigue.

A second issue concerns constraints on the driver's pacing of his work. It is clear that the ability to plan one's driving trips to avoid over extending oneself is important in preventing fatigue. The facility to take a rest break and perhaps a sleep or a nap is one of the most effective ways of counteracting fatigue. On the other hand, delays and hold-ups which interfere with the driver's mental projections of the trip ahead and which prevent destinations being easily attainable are experienced as distressing and frustrating. Indeed it could be said that one of the more potentially dangerous circumstances of professional driving is the frequent necessity to achieve a destination within a certain time despite delays which are not of the driver's control and despite the lack of availability of psychological or physiological resources for accomplishing that task. The issue of flexibility of scheduling of driving task demands is important not only in ensuring that the demands of the task are congruent with the daily and weekly demands of sleep, rest, food and domestic and social activities, but also in terms of pacing of effort within the working day. There may be a physiological cost involved in the disruption of ultradian rhythms of activation. Certainly tasks that are self-paced have been found to be less susceptible to disruption by noise, sleep loss and working at night. Hence the issue of control over pacing, and flexibility of scheduling of driving and work activities may be rather more important than the attention it has attracted would suggest.

A third issue concerns the extent to which it is possible to simulate conditions giving rise to fatigue, in the laboratory or in an artificially controlled study. By their very nature a laboratory or the accoutrements of behavioural or physiological research create a special situation to which the experimental subjects will respond in a way that may be very different from how they behave in their everyday lives, whatever their intentions concerning their role as subjects of the study, or perhaps even because of their intentions. The concept of motivation—specifically willingness or preparedness to continue activity—is so central to the concept of fatigue that it is hard to see how what a person is prepared to undergo in the (for the subject) rather novel, highly unusual and 'unnatural' circumstances of an experiment is comparable to what he will tolerate or undertake in his work as a matter of routine over a period of years. The problem has its origins in the two traditions of fatigue research—the one being more analytic of the nature of fatigue and theoretically sophisticated, but rather harder to test (associated with Bartley and Chute), the other being pragmatic and operational, regarding fatigue as the behavioural result of continued activity regardless of the context of that activity (originating with Bartlett).

The importance of context is highlighted by the fact that repeatedly it is the studies which most nearly approximate to normal conditions which have obtained results which are more predictable, reliable and interpretable in relation to the investigators' hypotheses. Fatigue, as Cameron has pointed out, is embedded in the whole pattern of life of those who experience it, and that should be the starting point for research into its origins and effects. This implies that the origins of fatigue are rather more varied than simply extended performance of one activity, but should, at the very least, include patterns of sleep and rest, and the balance between work, leisure and domestic demands and activities. Furthermore, the notion of fatigue expresses an aspect of the relationship between an individual and the activity or work he is performing. The attitudes, aspirations and motivation of that person in performing that activity and the social norms to which he adheres are therefore highly relevant to the way in which fatigue is expressed; the expression of fatigue may not therefore be entirely predictable from the objective conditions under which that activity is carried out. This is an aspect of fatigue which has been almost entirely neglected in studies of driving and drivers.

7.5. *Practical implications of research*

Despite inconsistencies and gaps in the evidence it is still possible to draw out some general practical implications of the research that has been reviewed. These are considered below.

Safety levels

Heavy goods vehicle driving is a dangerous occupation, and there is a need for a rather more effective monitoring of levels of safety in the industry. This should refer to the risks to the driver in terms of his time of exposure, rather than in terms of distance travelled which is a common way of interpreting accident statistics. Further investigation is urgently required into the role of fatigue in accidents, particularly in relation to the timing and duration of driving shifts. The evidence currently available indicates that such factors are significant in accident causation, but also leaves many unanswered questions concerning the parameters of these effects. The technical feasibility of doing such research is made much more favourable by the existence of automatic recording devices in truck cabs. The importance of this question is underlined by the high rates of death and serious injury inflicted on the general road using public who are involved in accidents with heavy goods vehicles.

Further research

Researchers are always calling for more research, and the magnitude of the questions that require an answer tend to make the hard evidence already accumulated by research seem insignificant by comparison. However, in calling for more and better research, three general points are worth making.

Firstly, finding answers to pragmatic and practical questions concerning fatigue and driving also requires considerable theoretical advance. It has been too often pointed out that one cannot simply measure 'fatigue'. Secondly, there is a need for a greater co-ordination of research effort. The more successful research programmes have been those which have been run on sufficient scale to encompass the range of variables, both independent and dependent, which are known to interact and whose inter-relationships cannot easily be ignored. There is thus an important role for a large scale programme of research with definite objectives, though it should not eclipse the importance of small independent theoretically innovative projects. And thirdly, research into fatigue and driving should be firmly rooted in the actual practices and experiences of drivers.

Regulating drivers' hours

Whatever regulations governing drivers' hours there are should be effectively enforced; they obviously are not in many countries. This tends to have the effect of putting the driver outside the law if he is to fulfil the requirements of his job, for the norms of working and driving hours will bear little relation to regulations that are not enforced. Furthermore, such regulations become an additional source of stress for the driver, for the risk of punitive sanctions is not entirely removed, rather than the regulations affording him protection against excessive hours. Regulations can only be effective if they do in fact set the boundaries to what can and does occur in practice.

Liability for punitive sanctions for contravention of the regulations should only fall on the driver in so far as he is responsible for determining the hours he works or drives, and to a large extent one suspects this responsibility is minimal.

It is difficult to make firm recommendations about the maximum number of hours of driving that should be permissible. The evidence concerning deterioration in driving performance is too difficult to interpret precisely in terms of safety, and the accident statistics do not provide an optimal cut-off point. It is clear that there is a relationship between driving hours and accidents in that a higher rate of accidents occurs later rather than earlier in trips. So it is impossible to say at this stage how much gain in safety would be achieved by reducing driving hours from 10 to eight, or from eight to less, for example.

However, what is more important than simply controlling hours of driving is the control of hours of work and rest and the distribution of these throughout the day and week. For most professional drivers their duties include far more than simply driving—time is spent on administration, maintenance, loading and waiting, for example. All these activities are work, and even waiting time, because it is usually unpredictably long, cannot be used effectively for rest. Time spent on these activities will affect the driver's state of tiredness or fatigue while he is driving, but also may have implications for his opportunities for rest and recuperation. It therefore seems highly inappropriate that while, in the EEC regulations, driving time is clearly and straightforwardly restricted, it is possible legally to work 84 hours per week (O'Hanlon 1979). There is clearly a strong need for the implementation of the proposed amendments to the EEC regulations specifying maximum daily spread-overs and weekly working times.

What no sets of regulations on drivers' hours take into acount are the difficulty of getting enough sleep during the day and the problems of driving during the night. Thus regulations should preserve for the driver sufficient time, at the *right* time, to get adequate rest and sleep and recognise that night work is particularly taxing.

What then should one conclude? What follows is inevitably fairly arbitrary, but then any regulation will be arbitrary. Concerning daytime work and driving, despite all the problems of interpretation the balance of evidence does favour the conclusions of Harris, Mackie and their colleagues that the congruence of changes in driving performance, and the physiological and subjective state of the driver, indicate that the current US regulations permit too long periods of work and driving, particularly over a 6-day week. Thus, curtailing the working day to the level of the EEC proposal (12 hours) or even the previous British level (11 hours) would be likely to increase safety, but it would be hard to predict by how much. Alternatively, the working week could be curtailed to the same effect to four or five days. An increase in safety levels may follow a curtailment of the driving day to, say, eight hours rather than 10, but it appears that controlling working hours is more important than controlling driving hours in preventing fatigue, and it may be desirable to preserve a greater flexibility of driving hours within a more firmly controlled boundary of working hours.

It should be noted that these comments apply only to those situations where there is strong evidence of fatigue; such changes might be expected to eliminate some of the problem of fatigue but not to eradicate it. Working norms which exceed those in other industries by a large margin would still be possible, thus one could still invoke grounds of the drivers' social well-being to argue for a further curtailment. It is worth noting that the International Labour Office Convention 67 on hours of work in road transport (which has not been widely ratified) stipulates a maximum working week of 48 hours, a maximum working day of eight hours (nine hours if the average for the week is eight), though allows for 75–100 hours per year of additional overtime (ILO 1977).

Concerning shiftworking and working at night, perhaps the clearest and most concise criteria for optimal shift systems are those given by Rutenfranz *et al.* (1977), which should be seriously considered in relation to the framing of any set of regulations or collective agreements on hours of work. Their main recommendations are as follows. Single night-shifts are better than consecutive night-shifts because they leave the circadian rhythm intact and do not involve continuous re-entrainment of rhythms between work days and rest days. At least 24 hours of free time should be allowed after each night-shift to permit proper sleep and recovery. The same requirements should hold for early morning shift times (which are very common in truck driving), and which also prevent the worker getting sufficient sleep the night before. The length of the work shift should not normally exceed eight hours and should depend on the nature of the work. Shift systems should preferably have short cycles and a regular rotation, to make it easier for the worker and his family to plan their social life, and continuous shift systems should ensure as many free weekends as possible, to permit participation in social and domestic activities.

Rutenfranz *et al.* also suggest that shiftworking is contra-indicated in the following groups: those under 25 or over 50 years of age (except in the latter case experienced and well adapted shiftworkers); those with a history of digestive disorders, diabetes, thyrotoxicosis or epilepsy; those who live alone; and those with inadequately soundproofed sleeping accommodation. They also emphasize the importance of thorough medical supervision of shiftworkers, particularly those beginning shiftwork and older workers. Because adaptation to shiftwork becomes more difficult with increasing age, in order to discourage shiftworkers from the temptation to continue shiftworking despite increasing problems of adaptation (because of financial commitments, of job security, etc.), it would be advantageous to have a right of transfer for shiftworkers to equivalent work on the day-shift (with retraining if necessary), or for provision to be made for early retirement for those who have worked shifts for a number of years.

Some may find these restrictions on shift systems to be a little severe, and it is undoubtedly true that many professional drivers find very little problem in coping with their shift system. However, the evidence of an association between driving at night and accident risk is very clear. While no shift system could be expected to entirely eliminate the problem of drowsiness at night, any system should be designed to ensure, at the very least, sufficient regular night-time sleep (which should preferably be taken at home, but sleep in the cab berth should not be considered a satisfactory substitute) and shorter working shifts; and that these criteria should apply to early morning shifts as well as night shifts.

Accident severity

There is evidence to suggest that truck drivers would benefit from a stronger cab structure which would afford better protection in the case of accident, and that, given this protection, seat belts would reduce the severity of injuries. The severity of injury to occupants of other vehicles who are in collision with trucks can also be reduced by measures to prevent these vehicles under-riding the truck or trailer, particularly the rear or sides; such measures are increasingly being applied.

It has been suggested that one way of reducing the number of accidents involving trucks and other vehicles (which tend to be very severe in terms of death or injury) is to attempt to separate the two types of traffic, and that an increase in shiftworking among truck drivers can be seen in this light (McDonald 1981). This would, however, increase the possibility of falling-asleep and single-vehicle accidents, which tend to carry the most risk of severe injury for the truck driver. Furthermore, the argument depends on a complete separation of night-time truck traffic and daytime other traffic with little or no overlap. The worst combination perhaps occurs with the overlap of early morning shifts with normal daytime traffic, where there is a combination of sleep deprivation, prolonged driving and high traffic density. So long as shift systems are organized to minimize the effects of sleep loss and prolonged driving at night, and so long as the night or early morning shift does not overlap with periods of high traffic density (which create additional driving stresses), night work may reduce the danger posed by trucks to other road users. Many drivers themselves, despite the disadvantages of working at night, find driving at night easier because of the clearer roads.

Truck cab environmental standards

There is a need for continued improvements in cab environmental standards, particularly in relation to noise, vibration and seating. Such improvements should not only contribute to a betterment in drivers' health (in terms of protecting hearing and reducing back problems), but also prevent adverse effects of these variables on the drivers' performance (e.g., attention and visual acuity) and reduce the experience of fatigue, discomfort, annoyance, etc.

It does not seem appropriate to consider factors which make for discomfort to be good methods for keeping the driver awake. The prevention of drowsiness and fatigue is better promoted by ensuring adequate rest and sleep and preventing excessive work.

Flexibility of schedules

There is plenty of evidence that people are usually sensitive to their own state of fatigue and drowsiness and their fitness to continue driving. Problems of fatigue therefore arise when drivers are constrained by professional or other commitments to continue driving despite feeling fatigued. One can only appeal to those who determine or negotiate driving schedules within the haulage industry to ensure that those schedules are sufficiently flexible to enable the driver to fulfil his work commitments (including a margin of error for delays) without overtaxing his abilities, or preventing him fulfilling his domestic and social needs, and allowing him to pace his own work throughout the day.

The needs of the driver in relation to scheduling and facilities for rest, food and hygiene, involve far more than the responsibilities of the employer, but involve facilities at the customer's depot, at customs posts, and en route, for example. The evidence suggests that drivers' needs are not sufficiently catered for in these aspects of his working conditions.

Norms of work in the haulage industry

Working norms within the road haulage industry diverge quite markedly from those in other industries. Hours of work are long, much time is spent away from home, and there is a tendency for domestic and social life to be restricted to the weekends. Perhaps one compensation for this virtual exclusion from the working week of everything except work and the reconstitution of labour power is the feeling of autonomy and freedom from managerial control which characterizes the driver's work.

It is clear also that for many drivers the pattern of work is changing, with an increasing intensification and supervision of work and the introduction of shiftwork. There are potential benefits in some of these changes in that reducing working hours will permit a relationship between work and leisure which is closer to that enjoyed by other workers, and prevent fatigue resulting from excessive hours of work. There are also potential disadvantages for the driver in the loss of control over his work (though the extent of this must be subject to some process of formal or informal negotiation), in the problems caused by shiftwork, and in the ever increasing mileages performed by some drivers (which may imply an increased risk of exposure to accidents). In this process of change it is important to try to ensure that in doing away with excessive work demands of one type, one does not merely substitute other countervailing pressures and dangers.

References

AAOO, 1959, *Guide for Conservation of Hearing in Noise* (Rochester, Minn.: American Academy of Opthalmology and Otolaryngology).

ACGIH, 1975, *Threshold Limit Value Data for 1975* (Cincinnati, Ohio: American Conference of Government Industrial Hygienists).

AANONSEN, A., 1964, *Shift Work and Health* (Oslo: Universitetsforlaget).

ADEY, W. R., 1969, Spectral analysis of EEG data from animals and man during alerting, orienting and discriminative responses, in *Attention in Neurophysiology*, Evans, C. R. and Mulholland, T. B. (eds.) (London: Butterworth).

ADUM, O., 1975, Shiftwork of professional drivers, in *Experimental Studies of Shiftwork*, Colquhoun, W. P. *et al.* (eds.) (Opladen: Westdeutscher Verlag).

DE ALMEIDA, A., 1950, Influence of electric punch card machines on the human ear. *Archives of Otolaryngology*, **51**, 215–222.

AN FORAS FORBARTHA, 1976, *Road Accident Facts 1975* (Dublin: An Foras Forbartha).

AN FORAS FORBARTHA, 1980, *Road Accident Facts 1979* (Dublin: An Foras Forbartha).

ANTICAGLIA, J. R. and COHEN, A., 1970, Extra-auditory effects of noise as a health hazard. *American Industrial Hygiene Association Journal*, **31**, 277–281.

ASCHOFF, J., HOFFMAN, K., POHL, H. and WEVER, R., 1975, Re-entrainment of circadian rhythms after phase shifts of the Zeitgeber. *Cronobiologica*, **2**, 23–80.

AUFFRET, R., SERIS, H., BERTHOZ, A. and FATRAS, B., 1967, Evaluation d'une tâche perceptive par la mesure de la variabilité du rythme cardiaque. Application à une tâche de pilotage. *Travail Humain*, **30**, 309–310.

AX, A. F., 1953, The physiological differentiation of fear and anger in humans. *Psychosomatic Medicine*, **15**, 433–442.

BABKOV, N. F., 1973, La perception de la route par le conducteur—nouveau facteur dans la conception des projets de routes, paper presented at the *First International Conference on Driver Behaviour*, Zurich, October 1973 (Courbevoie, France: IDBRA).

BAKER, C. H., 1960, Maintaining the level of vigilance by artificial signals. *Journal of Applied Psychology*, **44**, 336–338.

BAKER, J. S., 1967, Single vehicle accidents on Route 66. *Journal of Criminal Law, Criminology and Police Science,* **58,** 583–595.

BAKER, J. S., 1968, *Single Vehicle Accidents: a Summary of Research Findings* (Washington, D.C.: Automotive Safety Foundation).

BAKER, S. P., WONG, J. and MASEMORE, W. C., 1975, Fatal tractor-trailer crashes: considerations in setting relevant standards, in *Proceedings of the Fourth International Congress on Automotive Safety, July 14–16, 1975* (Washington, D.C.: US Government Printing Office), pp. 25–36.

BARBARIK, P., 1968, Automobile accidents and driver reaction patterns. *Journal of Applied Psychology,* **52,** 49–54.

BARTH, J. L., HOLDING, D. H. and STAMFORD, B. A. (undated), *Risk versus Effort in the Assessment of Motor Fatigue,* Department of Psychology and Exercise Physiology Laboratory, University of Louisville (unpublished).

BARTLETT, F. C., 1953, Psychological criteria for fatigue, in *Symposium on Fatigue,* Floyd, W. F. and Welford, A. T. (eds.) (London: H. K. Lewis).

BARTLEY, S. H. and CHUTE, E., 1947, *Fatigue and Impairment in Man* (New York: McGraw-Hill).

BARTON, J. C. and HEFNER, R. E., *Whole Body Vibration Levels: a Realistic Baseline for Standards.* SAE Paper no. 760415, Society of Automotive Engineers, Warrendale, Pa.

BELT, B. L., 1969, *Driver Eye Movements as a Function of Low Alcohol Concentrations.* Technical Report, Engineering Experiment Station, Ohio State University, Columbus, Ohio.

BJORKMAN, M., 1963, An exploratory study of predictive judgements in a traffic situation. *Scandinavian Journal of Psychology,* **4,** 65–76.

BOADLE, J., 1976, Vigilance and simulated night driving. *Ergonomics,* **19,** 217–225.

BÖCHER, W., 1975, *Untersuchungen über das Berufsbild des Lastkraftwagenfahrers* (Investigations relating to the professional delineation of the truck driver). Technischer Überwachungs-Verein Rheinland e.V. Medizinisch-Psychologisches Institut, 5 Köln 1, Postfach 101750.

BOGGS, D. H. and SIMON, J. R., 1968, Differential effect of noise on tasks of varying complexity. *Journal of Applied Psychology,* **52,** 148–153.

BOTTOM, C. G. and ASHWORTH, R., 1973, Driver behaviour at priority type intersections, paper SM7a presented at the *First International Conference on Driver Behaviour,* October 1973, Zurich (Courbevoie, France: IDBRA.)

BROADBENT, D. E., 1950, *The Twenty Dials Test under Quiet Conditions.* Medical Research Council Reports, Applied Psychology Unit, No. 130/50.

BROADBENT, D. E., 1951, *The Twenty Dials Test under Noise Conditions.* Medical Research Council Reports, Applied Psychology Unit, No. 160/51.

BROADBENT, D. E., 1953, Noise, paced performance and vigilance tasks. *British Journal of Psychology,* **44,** 295–303.

BROADBENT, D. E., 1954, Some effects of noise on visual performance. *Quarterly Journal of Experimental Psychology,* **6,** 1–5.

BROADBENT, D. E., 1971, *Decision and Stress* (London: Academic Press).

BROADBENT, D. E., 1978, The current state of noise research: reply to Poulton. *Psychological Bulletin,* **85,** 152–167.

BROADBENT, D. E. and GREGORY, M., 1965, Effects of noise and of signal rate upon vigilance analysed by means of decision theory. *Human Factors,* **7,** 155–162.

BROADBENT, D. E. and LITTLE, E. A. J., 1960, Effects of noise reduction in a work situation. *Journal of Occupational Psychology,* **34,** 133–140.

BROHM, F. and ZLAMAL, J., 1962, Noise in motor transport. *Casopis Lekaru Ceskyek,* **101,** 300–307.

BROWN, I. D., 1962a, Studies of component movements, consistency and spare capacity of car drivers. *Annals of Occupational Hygiene,* **5,** 131–143.

BROWN, I. D., 1962b, Measuring the spare mental capacity of car drivers by a subsidiary auditory task. *Ergonomics*, **5**, 247–250.

BROWN, I. D., 1964, The measurement of perceptual load and reserve capacity. *Transactions of the Association of Industrial Medical Officers*, **14**, 44–49.

BROWN, I. D., 1965, A comparison of two subsidiary tasks used to measure fatigue in car drivers. *Ergonomics*, **8**, 467–473.

BROWN, I. D., 1967a, Car driving and fatigue. *Triangle* (Sandoz Journal of Medical Science), **8**, 131–137.

BROWN, I. D., 1967b, Decrement of skill observed after 7 hrs of car driving. *Psychonomic Science*, **7**, 131–132.

BROWN, I. D., 1976, Personal communication, M.R.C. Applied Psychology Unit, Cambridge, U.K.

BROWN, I. D. and POULTON, E. C., 1961, Measuring the spare "mental capacity" of car drivers by a subsidiary task. *Ergonomics*, **4**, 35–40.

BROWN, I. D., SIMMONDS, D. C. V. and TICKNER, A. H., 1967, Measurement of control skills, vigilance and performance on a subsidiary task during 12 hours of car driving. *Ergonomics*, **10**, 665–673.

BROWN, I. D., TICKNER, A. H. and SIMMONDS, D. C. V., 1966, Effects of prolonged driving upon driving skill and performance of a subsidiary task. *Industrial Medicine and Surgery*, **35**, 760–765.

BROWN, I. D., TICKNER, A. H. and SIMMONDS, D. C. V., 1970, Effect of prolonged driving on overtaking criteria. *Ergonomics*, **13**, 239–242.

BROWN, J. D. and HUFFMAN, W. J., 1972, Psychophysiological measures of drivers under actual driving conditions. *Journal of Safety Research*, **4**, 172–178.

BRUUSGAARD, A., 1969, Shift work as an occupational health problem, in *On Night and Shift Work*, Swensson, A. (ed.) (Stockholm: Studia Laboris et Salutis), vol. 4, pp. 9–14.

BUCK, L., 1966, Reaction time as a measure of perceptual vigilance. *Psychological Bulletin*, **65**, 291–304.

BUGARD, P., SOUVRAS, H., VALADE, P., COSTE, E. and SALLE, J., 1953, Le syndrome de fatigue et les troubles auditifs des metteurs au point d'aviation. *La Semaine des Hospitaux*, **29**, 65–70.

BUIKHUISEN, W. and JONGMAN, R. W., 1972, Traffic perception under the influence of alcohol. *Quarterly Journal of Studies on Alcohol*, **33**, 800–806.

BUREAU OF MOTOR CARRIER SAFETY, 1971, *Physical Condition Report of Commercial Drivers Involved in Accidents for Year 1970* (Washington, D.C.: US Department of Transportation).

BURNS, N. M., BAKER, C. A., SIMONSON, E. and KEIPER, C., 1966, Electrocardiogram changes in prolonged automobile driving. *Perceptual and Motor Skills*, **23**, 210.

BURNS, W., 1973, *Noise and Man* (London: John Murray).

BURNS, W. and ROBINSON, D. W., 1970, *Hearing and Noise in Industry* (London: HMSO).

BURTON, A. C. and EDHOLM, O. G., 1955, *Man in a Cold Environment* (London: Arnold).

CAMERON, C., 1973, Fatigue and driving—a theoretical analysis. *Australian Road Research*, **5**, 36–44.

CAMERON, C., 1974, A theory of fatigue, in *Man Under Stress*, Welford, A. T. (ed.) (London: Taylor & Francis).

CARLSON, L. D. and HSIEH, A. C. L., 1965, Cold, in *The Physiology of Human Survival*, Edholm, O. G. and Bacharach, A. L. (eds.) (New York: Academic Press).

CARPENTIER, J. and CAZAMIAN, P., 1977, *Night Work* (Geneva: International Labour Office).

CARRÉ, J. R. and HAMELIN, P., 1978, *Rapports Sociaux de Production, Mode de Régulation et Conditions de Travail des Conducteurs Routiers*. Organisme National de Sécurité Routière, Montlhèry.

CHATFIELD, B. U. and HOSEA, H. R., 1972, Fatal accidents on the interstate system 1968–1971. *Public Roads,* **37,** 108–118.

CHENCHANNA, P. and KRISANANKUTTY, U. K., 1969, Driving performance of drivers under vertical mechanical vibrations. *Automobilismo e Automobilismo Industriale* (Rome), No. 9–10, 619–638.

CLAYTON, A. B., 1972, An accident-based analysis of road-user errors. *Journal of Safety Research,* **4,** 69–74.

CLOSE, W. S. and CLARKE, R. M., 1972, *Truck noise II: Interior and Exterior A-weighted Sound Levels of Typical Highway Trucks.* US Department of Transportation Report OST/TST-72-2 Z (Washington, D.C.: US Department of Transportation).

COERMANN, R. R., 1961, *The Mechanical Impedance of the Human Body in Sitting and Standing Positions at Low Frequencies.* ASD Technical Report 61-492, Aeronautical Systems Division Aerospace Medical Laboratory, Wright-Patterson Air Force Base, Ohio.

COHEN, A., 1973, Extra-auditory effects of noise. *National Safety News (U.S.),* **108,** 93–99.

COHEN, A., HUMMEL, W. F., TURNER, J. W. and DUVER-DUBOS, F. N., 1966, *Effects of Noise on Task Performance.* Report RR-4, US Department of Health, Education and Welfare.

COHEN, A. S., 1978, *Eyemovements Behaviour while Driving a Car: a Review.* Report TR-78-TH 4, Swiss Federal Institute of Technology, Zurich.

COHEN, A. S. and HIRSIG, R., 1970, Causality between drivers' successive eye fixations. *Perceptual and Motor Skills,* **48,** 974.

COHEN, J., 1962, Alcohol and road accidents. *Triangle* (Sandoz Journal of Medical Science), **5,** 214–220.

COHEN, J., DEARNALEY, E. J. and HANSEL, C. E. M., 1956, Risk and hazard. *Operational Research Quarterly,* **7,** 67–82.

COHEN, J., DEARNALEY, E. J. and HANSEL, C. E. M., 1958, The risk taken in driving under the influence of alcohol. *British Medical Journal,* 1438–1442.

COLQUHOUN, W. P., 1971, Circadian variations in mental efficiency, in *Biological Rhythms and Human Performance,* Colquhoun, W. P. (ed.) (London: Academic Press).

COLQUHOUN, W. P., 1982, Biological rhythms and performance, in *Biological Rhythms, Sleep, and Performance,* Webb, W. B. (ed.) (Chichester: John Wiley), pp. 59–86.

COLQUHOUN, W. P. and FOLKARD, S., 1978, Personality differences in body temperature rhythm, and their relation to its adjustment to night work. *Ergonomics,* **21,** 811–817.

COLQUHOUN, W. P., FOLKARD, S., KNAUTH, P. and RUTENFRANZ, J. (eds.), 1975, *Experimental Studies of Shiftwork* (Opladen: Westdeutscher Verlag).

COMMISSION OF THE EUROPEAN COMMUNITIES, 1976, *Proposal for a Council Regulation on the Harmonization of Certain Social Legislation Relating to Road Transport.* Com (76) 85 final, Brussels, 3 March 1976.

COMMISSION OF THE EUROPEAN COMMUNITIES, 1980, *Seventh Report by the Commission to the Council on the Implementation of Council Regulation (EEC) no. 543/69 of 25 March 1969 on the Harmonization of Certain Social Legislation Relating to Road Transport.* COM (80) 486 final, Commission of the European Communities, Brussels.

CONNOLLY, P. L., 1966, *Visual Considerations of Man, the Vehicle and the Highway. Part II.* SAE Report SP 279, Society of Automotive Engineers, New York.

CONROY, R. T. W. L. and MILLS, J. N., 1969, Circadian rhythms and shift work, in *On Night and Shift Work,* Swensson, A. (ed.) (Stockholm: Studia Laboris et Salutis), vol. 4, pp. 42–46.

CORCORAN, D. W. J., 1962, Noise and loss of sleep. *Quarterly Journal of Experimental Psychology,* **14,** 178–182.

COX, T., 1978, *Stress* (London: Macmillan).

CRAWFORD, A., 1961, Fatigue and driving. *Ergonomics,* **4,** 143–154.

CRAWFORD, A., 1963, The overtaking driver. *Ergonomics,* **6,** 153–170.

CREUTZFELDT, O., GRÜNEWALD, I., SIMONOVNA, I. and SCHMITZ, H., 1969, Changes in the basic rhythms of the EEG during performance of mental and visuomotor tasks, in *Attention in Neurophysiology,* Evans, C. R. and Mulholland, T. B. (eds.) (London: Butterworths).

CROWLEY, F. and HEARNE, R., 1972, *The Distribution of Accidents by Hour of Day in the Republic of Ireland.* Report RS 98 (Dublin: An Foras Forbatha Teo).

CULLEN, J., FULLER, R. and DOLPHIN, C., 1979, Endocrine stress responses of drivers in a "real-life" heavy goods vehicle driving task. *Psychoneuroendocrinology,* **4,** 107–115.

CULLEN, J. H., 1976, Psycho-biological aspects of fatigue, in *Workshop on the Influence of Driver Fatigue on Traffic Accidents,* Dourdan, Sep. 1976 (Montlhèry, France: L'Organisme National de Sécurité Routière).

CURRIE, L., 1969, The perception of danger in a simulated driving task. *Ergonomics,* **12,** 841–849.

CURRY, G. A., HIEATT, D. J. and WILDE, G. J. S., 1975, *Task Load in the Motor Vehicle Operator: A Comparative Study of Assessment Procedures.* Report CR 7504, Department of Psychology, Queen's University, Kingston, Ontario (not published).

DAVIES, D. R., 1968, Physiological and psychological effects of exposure to high intensity noise. *Applied Acoustics,* **1,** 215–233.

DAVIES, D. R. and KROVIC, A., 1965, Skin conductance, alpha-activity and vigilance. *American Journal of Psychology,* **78,** 304–306.

DE KOCK, A. R., 1968, *Relationship between Decision Making under Conditions of Risk and Selected Psychological Tests.* Technical Report No. 9, Human Factors Laboratory, Department of Psychology, University of South Dakota (not published).

DENTON, D. G., 1967, *The Effect of Speed and Speed Change on Drivers Speed Judgement.* Report LR 97, Transport and Road Research Laboratory, Crowthorne, Berks.

DENTON, G. G., 1968, The use made of a speedometer as an aid in driving. *Ergonomics,* **12,** 447–452.

DEPARTMENT OF EMPLOYMENT, 1979, *New Earnings Survey 1978* (London: HMSO).

DEPARTMENT OF THE ENVIRONMENT, HEAVY GOODS VEHICLE LICENSING SECTION, 1980, personal communication.

DEPARTMENT OF TRANSPORT, 1980, *Transport Statistics Great Britain 1969–1979* (London: HMSO).

DEPARTMENT OF TRANSPORT, 1981, *Transport Statistics Great Britain 1970–1980* (London: HMSO).

DEPARTMENT OF TRANSPORT, 1982, *Road Accidents, Great Britain 1981* (London: HMSO).

DERMER, M. and BERSCHEID, E., 1972, Self-report of arousal as an indicant of activation level. *Behavioural Science,* **17,** 420–429.

DOBBINS, D. A., TIEDEMANN, J. G. and SKORDAHL, D. M., 1963, Vigilance under highway driving conditions. *Perceptual and Motor Skills,* **16,** 38.

DONNERSTEIN, E. and WILSON, D. W., 1976, Effects of noise and perceived control on ongoing and subsequent aggressive behaviour. *Journal of Personality and Social Psychology,* **34,** 744–781.

DRETTNER, B., HEDSTRAND, H. A. and KLOCKHOFF, I., 1975, Cardiovascular risk factors and hearing loss—a study of 3,000 fifty year old men. *Acta Otolaryngolica,* **79,** 366–371.

DRORY, A. and SHINAR, D., 1982, The effects of roadway environment and fatigue on sign perception. *Journal of Safety Research,* **13,** 25–32.

DUFFY, E., 1957, The psychological significance of the concept of 'arousal' or activation. *Psychological Review,* **64,** 265–275.

DUFFY, E., 1962, *Activation and Behaviour* (New York: Wiley).

DUMKINA, G. Z., 1966, *Some Clinical and Physiological Studies of Workers Subjected to Stable Noise*. Army Foreign Science and Technology Center, Charlottesville, Virginia, Report No. AD-722 449/5WJ, 1973.

DUPUIS, H., 1965, Fortlaufende Pulsfrequenzschreibung bei Kraftfahren und ihre Interpretation, in Supplement 3 to *Arbeitswissenschaft* (Mainz: Krausskopf-Verlag).

DUPUIS, H. and HARTUNG, E., 1968, Untersuchung über die Fahrerplatzgestaltung für den "Standard-Linienbus" und Einmannbetreib, *ATZ Automobiltechnische Zeitschrift*, **70**, No. 11 (Stuttgart: Franckh'schse Verlagshandlung).

DUREMAN, E. I. and BODEN, C., 1972, Fatigue in simulated car driving. *Ergonomics*, **15**, 299–308.

DYOS, H. J. and ALDCROFT, D. H., 1969, *British Transport* (Leicester: University Press).

EDMONDSON, J. L. and OLDMAN, M., 1974, *An Interview Study of Heavy Goods Vehicle Drivers*. Contract report 74/14, Institute of Sound and Vibration Research, University of Southampton (not published).

EDWARDS, D. S., HAHN, C. P. and FLEISHMAN, E. A., 1969, *Evaluation of Laboratory Methods for the Study of Driver Behaviour*. Report No. R69-7, pp. 1–43, American Institute for Research (not published).

EDWARDS, W., 1968, Information processing, decision making and highway safety. Paper presented at *Second Annual Traffic Safety Research Symposium of the Automobile Insurance Industry* (not published).

EGELUND, N., 1982, Spectral analysis of heart rate variability as an indicator of driver fatigue. *Ergonomics*, **25**, 663–672.

ELLINGSTAD, V. S. and HEIMSTRA, N. W., 1970, Performance changes during the operation of a complex psychomotor task. *Ergonomics*, **13**, 693–705.

EMME, J., 1970, Analysis of truck noise during actual operation. *Acoustic Laboratory Report*, H. L. Blackford Inc., Orange, California (not published).

EVANS, L., 1970, Speed estimation from a moving automobile. *Ergonomics*, **13**, 219–230.

EVANS, L. and ROTHERY, R., 1974, Detection of the sign of relative motion when following a vehicle. *Human Factors*, **16**, 161–173.

EVANS, M. J., 1972, Infrasonic effects on the human organs of equilibrium. *Proceedings of the British Acoustical Society*, **1**, 71–104.

EVANS, M. J. and TEMPEST, W., 1972, Some effects of infrasonic noise in transportation. *Journal of Sound and Vibration*, **22**, 19–24.

FANGER, P. O., 1967, Calculation of thermal comfort: Introduction of a basic comfort equation. *ASHRAE Transactions*, **73**, II, 1–20.

FARACHI, B., DRAGONETTI, M. and ANNESI, A., 1971, Sicurezza in autostrada. Tamponamenti: diagnosi di un incidente. *Autostrade*, **13**(8), 44–55.

FELTON, J. R., 1978, The inherent structure, behaviour and performance of the motor freight industry. *Rivista Internazionale Di Economia Dei Transport*, **1**, 23–35.

FIRTH, P. A., 1973, Psychological factors influencing the relationship between cardiac arrhythmia and mental load. *Ergonomics*, **16**, 5–16.

FISHBEIN, W. I. and SALTER, L. C., 1950, The relationship between truck and tractor driving and disorders of the spine and supporting structures. *Industrial Medicine and Surgery*, **19**, 444–445.

FLOYD, W. F. and SANDOVER, J., 1972, Problems associated with the application of the results of vibration research to practical situations. Paper presented at the *Society of Environmental Engineers, Vibrations and Man Symposium*, April 1972 (not published).

FOLKARD, S., 1975, The nature of diurnal variations in performance and their implications for shift work studies, in *Experimental Studies of Shiftwork*, Colquhoun, W. P., Folkard, S., Knauth, P. and Rutenfranz, J. (eds.) (Opladen: Westdeutscher Verlag).

FOLKARD, S., 1981, Shiftwork and performance, in *Biological Rhythms, Sleep and Shiftwork*, Johnson, L. C., Tepas, D. I., Colquhoun, W. P. and Colligan, M. J. (eds.) (Lancaster: MTP Press), pp. 283–305.

FOLKARD, S., MONK, J. and LOBBAN, M. C., 1979, Towards a predictive test of adjustment to shiftwork. *Ergonomics*, **22**, 79–91.

FORBES, T. W. and MATSON, T. M., 1939, Driver judgements in passing on the highway. *Journal of Psychology*, **8**, 3–11.

FORD, B., 1971, Who is the sleepy driver? *Traffic Safety, Chicago*, Nov. 1971, 12–15.

FORET, J. and LANTIN, G., 1972, The sleep of train drivers: An example of the effects of irregular work schedules on sleep, in *Aspects of Human Efficiency*, Colquhoun, W. P. (ed.) (London: English Universities Press).

FOSTER, A. W., 1978, *A Truck Heavy Suspension for Improved Ride*. SAE Paper no. 780408, Society of Automotive Engineers, Warrendale, Pa.

FOX, R. H., 1965, Heat, in *The Physiology of Human Survival*, Edholm, O. G. and Bacharach, A. L. (eds.) (London: Academic Press).

FRANKENHAUSER, M., 1975, Sympathetic-adrenomedullary activity, behaviour and the psychosocial environment, in *Research in Psychophysiology*, Venables, P. and Christie, M. (eds.) (London: Wiley), pp. 71–94.

FRANKENHAUSER, M., 1981, Coping with job stress—a psychobiological approach, in *Man and Working Life*, Gardell, B. and Johansson, G. (eds.) (New York: Wiley), pp. 215–233.

FRIEDMAN, S. R., 1978, Changes in the trucking industry and the teamsters union: the Bonapartism of Jimmy Hoffa. *Insurgent Sociologist*, **8**(2 & 3), 56–62.

FRYMOYER, J. W., POPE, M. H., COVANZA, M. C., ROSEN, J. C., GOGGIN, J. E. and WILDER, D. G., 1980, Epidemiologic studies of low-back pain. *Spine*, **5**, 419–423.

FULLER, R. G. C., 1978, *Effect of Prolonged Driving on Heavy Goods Vehicle Driving Performance*. Report to US Army Research Institute for the Behavioral and Social Sciences, Alexandria, Virginia.

FULLER, R. G. C., 1980, Time headway in different following manoeuvres. *Perceptual and Motor Skills*, **50**, 1057–1058.

FULLER, R. G. C., 1981, Determinants of time headway adopted by truck drivers. *Ergonomics*, **24**, 463–474.

FULLER, R. G. C., 1983, *Prolonged Heavy Vehicle Driving Performance: Effects of Unpredictable Shift Onset and Duration and Convoy vs. Independent Driving Conditions*. European Research Office, U.S. Army, London.

FULLER, R. G. C., JOHNSTON, D., MATTHEWS, W. and McDONALD, N., 1975, *Smoking Motivation: The Heart Rate Hypothesis*. Department of Psychology, Trinity College, Dublin (not published).

GARDELL, B., ARONSSON, G. and BARKLOFF, K., 1981, *The Working Environment for Local Transit Personnel*. Dept. of Psychology, Univ. of Stockholm.

GEEN, R. G. and O'NEAL, E. C., 1969, Activation of cue-elicited aggression by general arousal. *Journal of Personality and Social Psychology*, **11**, 289–292.

GIBBS, W. L., 1968, Driver gap acceptance at intersections. *Journal of Applied Psychology*, **52**, 200–204.

GIBSON, J. J., 1950, *The Perception of the Visual World* (Boston: Houghton Mifflin).

GIBSON, J. J., 1954, Visual perception of objective motion and subjective movement. *Psychological Review*, **61**, 304–314.

GIBSON, J. J. and CROOKS, L. G., 1938, A theoretical field analysis of automobile driving. *American Journal of Psychology*, **51**, 453–471.

GIBSON, J. J., OLUM, P. and ROSENBLATT, F., 1955, Motion parallax and motion perspective in aircraft landings. *American Journal of Psychology*, **68**, 372–385.

VON GIERKE, H. E., 1975, *The ISO Standard Guide for the Evaluation of Human Exposure to Whole-Body Vibration*. SAE Paper no. 751009, Society of Automotive Engineers, Warrendale, Pa.

GISSANE, W. and BULL, J. P., 1973, Fatal car occupant injuries after car/lorry collisions. *British Medical Journal*, no. 1, 67–71.

GLASS, D. C. and SINGER, J. E., 1972, *Urban Stress: Experiments on Noise and Social Stressors* (New York and London: Academic Press).

GORDON, D. A., 1966, Experimental isolation of drivers' visual input. *Public Roads*, **33**, 266–273.

GORDON, D. A. and MAST, J. M., 1968, Drivers' decisions in overtaking and passing. *Public Roads*, **35**, 97–101.

GORDON, D. A. and MICHAELS, R. M., 1965, Static and dynamic visual fields in vehicular guidance. *Highway Research Record*, **84**, 1–15.

GRANDJEAN, E., WOTZKA, G., SCHAAD, R. and GILGEN, A., 1971, Fatigue and stress in air traffic controllers, in *Methodology in Human Fatigue Assessment*, Hashimoto, K., Kogi, K. and Grandjean, E. (eds.) (London: Taylor & Francis).

GRATTAN, E. and HOBBS, J. A., 1978, *Injuries to Occupants of Heavy Goods Vehicles*. Report LR 854, Transport and Road Research Laboratory, Crowthorne, Berks.

GREENSHIELDS, B. D., 1966, Changes in driver performance with time in driving. *Highway Research Record*, **122**, 75–88.

GRETHER, W. F., 1971, Vibration and human performance. *Human Factors*, **13**, 203–216.

GRETHER, W. F., HARRIS, C. S., MOHR, G. C., NIXON, C. W., OHLBAUM, M., SOMMER, H. C., THALER, V. H. and VEGHTE, J. H., 1971, Effects of combined heat, noise and vibration stress on human performance and physiological functions. *Aerospace Medicine*, **42**, 1092–1097.

GRIFFIN, M. J. and LEWIS, C. H., 1978, A review of the effects of vibration on visual acuity and continuous manual control, Pt. I Visual acuity. *Journal of Sound and Vibration*, **56**, 383–413.

GRIME, G. and HUTCHINSON, T. P., 1979, Vehicle mass and driver injury. *Ergonomics*, **22**, 93–104.

GRUBB, T. C., 1970, Narcolepsy, the sleep you can't control. *Traffic Safety, Chicago*, **70** (May), 8–36.

GRUBER, G. J., 1976, *Relationships between Wholebody Vibration and Morbidity Patterns among Interstate Truck Drivers*. DHEW/PUB/NIOSH-77-167, PB-275 610, National Technical Information Service, Springfield, Va.

GRUBER, G. J. and ZIPERMAN, H. H., 1974, *Relationship between Whole-body Vibration and Morbidity Patterns among Motor Coach Operators*. Contract no. HSM-00-72-047, US Department of Health Education & Welfare, Public Health Service, Center for Disease Control, National Institute for Occupational Safety and Health, Cincinnati, Ohio 45202.

GUIGNARD, J. C. and KING, P. F., 1972, Aeromedical aspects of vibration and noise, *AGARDograph No. 151*, NATO Advisory Group for Aerospace Research and Development, Neuilly sur Seine, France.

GWILLIAM, K. M. and MACKIE, P. J., 1975, *Economics and Transport Policy* (London: George Allen & Unwin).

HADDON, W., 1971, *Reducing Truck and Bus Losses—Neglected Countermeasures*. SAE Paper no. 710409, Society of Automotive Engineers, New York.

HAHN, W. W., 1973, Attention and heart rate: a critical appraisal of the hypothesis of Lacey and Lacey. *Psychological Bulletin*, **79**, 59.

HALE, A. R. and HALE, M., 1972, *A Review of the Industrial Accident Research Literature.* Research Paper, Committee on Safety and Health at Work (London: HMSO).

HAMELIN, P., 1975, *Conditions de Travail des Conducteurs Professionnels et Sécurité Routière* (Montlhèry, France: Organisme National de Sécurité Routière).

HAMELIN, P., 1980, *Les Conditions temporelles de Travail des Conducteurs Routiers et la Sécurité Routière* (Montlhèry, France: Organisme National de Sécurité Routière).

HANES, R. M., 1970, *Human Sensitivity to Whole-body Vibration in Urban Transportation Systems.* Report APL/JHU, TPR, 004, Applied Physics Laboratory, Johns Hopkins University, Silver Spring, Maryland.

HARRÉ, R. and SECORD, P. F., 1972, *The Explanation of Social Behaviour* (Oxford: Blackwell).

HARRIS, C. S., SOMMER, H. C. and JOHNSON, D. L., 1976, Review of the effects of infrasound on man. *Aviation, Space, and Environmental Medicine*, **47**, 430–434.

HARRIS, W., 1977, Fatigue, circadian rhythm and truck accidents, in *Vigilance: Theory, Operational Performance and Physiological Correlates*, Mackie, R. R. (ed.) (New York: Plenum).

HARRIS, W., MACKIE, R. R., ABRAMS, C., BUCKNER, D. N., HARABEDIN, A., O'HANLON, J. F. and STARKS, J. R., 1972, *A Study of the Relationships among Fatigue, Hours of Service and Safety of Operations of Truck and Bus Drivers.* Report 1727-2. PB 213, 963, Human Factors Research Inc., Santa Barbara Research Park, Goleta, California.

HART, P. E., 1959, The restriction of road haulage. *Scottish Journal of Political Economy*, **6**, 116–138.

HASHIMOTO, K., 1967, *Physiological Strain of High Speed Bus Driving on the Mei-Shin Expressway and the Effects of Moderation of Speed Restrictions.* Railway Labour Science Research Institute, Japan National Railways, Tokyo.

HEIMSTRA, N. W., 1970, The effects of 'stress fatigue' on performance in a simulated driving situation. *Ergonomics*, **13**, 209–218.

HELANDER, M., 1976, *Drivers' Reactions to Road Conditions—a Psychophysiological Approach.* Institutionen for Vabyggnad, Chalmers University of Technology, Goteborg, Sweden (not published).

HELANDER, M., 1978, Applicability of driver's electrodermal response to the design of the traffic environment. *Journal of Applied Psychology*, **63**, 481–488.

HENDERSON, P., 1977, *Road Accidents and Commercial Vehicle Drivers.* Department of Psychology, The Queen's University of Belfast (not published).

HERBERT, M. J., 1963, *Analysis of a Complex Skill: Vehicle Driving.* US Army Medical Research Lab. Report, 581 (Fort Knox, KY: US Army Medical Research Lab.).

HERBERT, M. J. and JAYNES, W. E., 1964, Performance decrement in vehicle driving. *Journal of Engineering Psychology*, **3**, 1–8.

HERIDER, E. A. and LE FEVRE, W. F., 1967, *The Truck Seating Art—its State and Future.* SAE Paper no. 670043, Society of Automotive Engineers, New York.

HEWLAND, RUDER & FINN INTERNATIONAL LTD., 1976, *Safety and Comfort Factors in Heavy Vehicle Driving.* A report for the Bostrom Division, UOP Ltd, HR& F, 15 Albermarle Street, London (not published).

HILDEBRANDT, G., ROHMERT, W. and RUTENFRANZ, J., 1974, 12 and 24 hr. rhythms in error frequency of locomotive drivers and the influence of tiredness. *International Journal of Chronobiology*, **2**, 175–180.

HILDEBRANDT, G., ROHMERT, W. and RUTENFRANZ, J., 1975, The influence of fatigue and rest period on the circadian variation of error frequency in shift workers (engine drivers), in *Experimental Studies of Shiftwork*, Colquhoun, W. P., Folkard, S., Knauth, P. and Rutenfranz, J. (eds.) (Opladen: Westdeutscher Verlag).

HILLS, B. L., 1980, Vision, visibility, and perception in driving. *Perception*, **9**, 183–216.

HOBSBAWM, E. J., 1964, *Labouring Men* (London: Weidenfeld and Nicolson).

HOCKEY, G. R. J., 1970, Effect of loud noise on attentional selectivity. *Quarterly Journal of Experimental Psychology*, **22**, 28–36.

HOCKEY, G. R. J., 1973, Changes in information-selection patterns in multisource monitoring as a function of induced arousal shifts. *Journal of Experimental Psychology*, **101**, 35–42.

HOFFMAN, H. and SCHNEIDER, B., 1967, Belastung, Leistungsgrenzen und Ermüdung bei Kraftfahrern. *Hefte zur Unfallheilkunde*, no. 91, pp. 113–122 (Berlin: Springer Verlag).

HÖGSTRÖM, K. and SVENSON, L., 1980, *Accidents Involving Volvo Trucks Resulting in Personal Injury*. Volvo Truck Corporation, Gothenburg.

HOLLOWELL, P. G., 1968, *The Lorry Driver* (London: Routledge and Kegan Paul).

HOME OFFICE, 1975, *Offences Relating to Motor Vehicles 1974* (London: HMSO).

HOOD, R. A., LEVENTHALL, H. G. and KYRIAKIDES, K., 1972, Some subjective effects of infrasound. *Proceedings of the British Acoustical Society*, **1**(3), Paper 71.107.

HORNE, J. A. and OSTBERG, O., 1976, A self-assessment questionnaire to determine morningness-eveningness in human circadian rhythm. *International Journal of Chronobiology*, **4**, 97–110.

HORNICK, R. J., 1962, The effects of whole-body vibration in three directions on performance. *Journal of Engineering Psychology*, **1**, 93–101.

HORVATH, S. M., DAHMS, T. and O'HANLON, J. F., 1971, Carbon monoxide and human vigilance: a deleterious effect of present urban concentrations. *Archives of Environmental Health*, **23**, 343–347.

HOSEA, H. R. and CHATFIELD, B. V., 1972, Characteristics of rural and urban fatal accidents on the Interstate Highway system, 1969–70. *Public Roads*, **37**, 22–28, US Department of Transportation.

HOWELL, K. and MARTIN, A. M., 1976, An investigation of the effects of hearing protectors on vocal communication in noise. *Journal of Sound and Vibration*, **41**, 181–196.

HUETING, J. F. and SARPHATI, H. R., 1966, Measuring fatigue. *Journal of Applied Psychology*, **50**, 535–538.

HULBERT, S. F., 1957, Drivers GSRs in traffic. *Perceptual and Motor Skills*, **7**, 305–315.

HURST, P. M., 1964, *Errors in Driver Risk Taking*. Division of Highway Studies, Institute for Research, State College, Pennsylvania (not published).

HURST, P. M., PERCHONOK, K. and SEGUIN, E. L., 1968, Vehicle kinematics and gap acceptance. *Journal of Applied Psychology*, **52**, 321–324.

HUTTON, J. D., 1972, *How Loud are Diesel Truck Cabs?* SAE Paper no. 720698, Society of Automotive Engineers, New York.

ILO, 1977, *Hours of Work and Rest Periods in Road Transport* (Geneva: International Labour Office).

ISO, 1971, *Assessment of Occupational Noise Exposure for Hearing Conservation Purposes*. Report no. R. 1999 (Geneva: International Organization for Standardization).

ISO, 1978, *Guide for the Evaluation of Human Exposure to Whole-Body Vibration, Draft International Standard ISO 2631-1978E* (Geneva: International Organization for Standardization).

IRVING, A., 1973, The perceptual problems of the driver, paper presented at the *First International Conference on Driver Behaviour*, October 1973, Zurich (Courbevoie, France: IDBRA).

JACKSON, K. F., 1956, *Aircrew Fatigue in Long Range Maritime Reconnaissance*. Air Ministry Flying Personnel Research Committee Report FPRC 907.2.

JANEWAY, R. N., 1975, *Human Vibration Tolerance Criteria and Applications to Ride Evaluation*. SAE Paper no. 750166, Society of Automotive Engineers, Warrendale, Pa.

JANSEN, G., 1961, Adverse effects of noise in iron and steel workers. *Stahl und Eisen*, **81**, 217–220.

JANSEN, G., 1969, *Effects of Noise on Physiological State*. Paper presented at the National Conference on Noise as a Public Health Hazard, American Speech and Hearing Association, Washington, D.C. February, 1969.

JANSSEN, W. H., MICHON, J. A. and HARVEY, L. O., 1976, The perception of lead vehicle movement in darkness. *Accident Analysis and Prevention*, **8**, 151–166.

JERISON, H. J., 1967a, Activation and long term performance, in *Attention and Performance*, Sanders, A. F. (ed.) (Amsterdam: North Holland).

JERISON, H. J., 1967b, Signal detection theory in the analysis of human vigilance. *Human Factors*, **9**, 285–288.

JEX, H. R., ZELLNER, J. W., JOHNSON, W. A. and KLEIN, R. H., 1981, *Comprehensive Measurement of Ride of In-Service Trucks*. SAE Paper no. 810045, Society of Automotive Engineers, Warrendale, Pa.

JOHANSSON, G., BERGSTRÖM, S., JANSSON, G., OTTANDER, C. and ORNBERG, G., 1963, *Drivers and Road Signs I*. Report No. 11, Department of Psychology, University of Uppsala, Sweden, November 1963.

JOHANSSON, G. and BACKLUND, F., 1970, Drivers and road signs. *Ergonomics*, **13**, 749–759.

JOHANSSON, G. and RUMAR, K., 1966, Drivers and road signs: A preliminary investigation of the capacity of drivers to get information from road signs. *Ergonomics*, **9**, 57–62.

JOHANNESSEN, H. G., 1970, *Seat Belts for Truck Occupant Safety*. SAE Paper no. 700347, Society of Automotive Engineers, New York.

JOHNSON, H. M., 1945, Index numerology and measures of impairment. *American Journal of Psychology*, **56**, 551–558.

JOHNSON, L. C., 1982, Sleep deprivation and performance, in *Biological Rhythms, Sleep and Performance*, Webb, W. B. (ed.) (Chichester: Wiley), pp. 111–141.

JOHNSON, L. C., TEPAS, D. I., COLQUHOUN, W. P. and COLLIGAN, M. J. (eds.), 1982, *Biological Rhythms, Sleep and Shiftwork* (Lancaster: MTP Press).

JOHNSSON, H. N. A. and HANSSON, L., 1977, Prolonged exposure to a stressful stimulus (noise) as a cause of raised blood pressure in man. *Lancet*, 1 (8002), 86–87.

JONES, B. F., FLYNN, R. H. and HAMMOND, G. C., 1941, Fatigue and hours of service of interstate truck drivers. *Public Health Bulletin*, 265 (Washington D.C.: US Government Printing Office).

JONES, D. M. and BROADBENT, D. E., 1979, Side effects of interferences with speech by noise. *Ergonomics*, **22**, 1073–1081.

JONES, H. V. and HEIMSTRA, N. W., 1964, Ability of drivers to make critical passing judgements. *Journal of Engineering Psychology*, **3**, 117–122.

JONES, I. S., 1976, *The Effects of Vehicle Characteristics on Road Accidents* (Oxford: Pergamon).

KAHNEMAN, D., 1973, *Attention and Effort* (New York: Prentice Hall).

KALSBEEK, J. W. H., 1971, Sinus arrhythmia and the dual task method in measuring mental load, in *Measurement of Man at Work*, Singleton, W. T., Fox, F. G. and Whitfield, D. (eds.) (London: Taylor & Francis).

KALSBEEK, J. W. H., 1973, Do you believe in sinus arrhythmia? *Ergonomics*, **16**, 99–104.

KALSBEEK, J. W. H. and ETTEMA, J. H., 1963, Continuous recording of heart-rate and the measurement of perceptual load. *Ergonomics*, **6**, 306–307.

KALSBEEK, J. W. H. and SYKES, R. N., 1967, Objective measurement of mental load. *Acta Psychologica*, **27**, 253–261.

KALUGER, N. A. and SMITH, G. L., JR., 1970, Driver eye movement patterns under conditions of prolonged driving and sleep deprivation. *Highway Research Record*, **336**, 92–106.

KAM, J.K.-H., 1980, Noise exposure levels among 20 selected truck drivers. *Journal of Environmental Health*, **43**, 83–85.

KAO, H. S., 1969, Feedback concepts of driver behaviour and the highway information system. *Accident Analysis and Prevention*, **1**, 65–76.

KASHIWAGI, S., 1971, Psychological rating of human fatigue, in *Methodology in Human Fatigue Assessment*, Hashimoto, K., Kogi, K. and Grandjean, E. (eds.) (London: Taylor & Francis).

KELSEY, J. L. and HARDY, R. J., 1975, Driving of motor vehicles as a risk factor for acute herniated lumbar intervertebral disc. *American Journal of Epidemiology*, **102**, 63–73.

KISHIDA, K., 1981, Subsidiary behaviour of truck drivers in rear-end collisions. *Journal of Human Ergology*, **10**, 113.

KLEITMAN, N. and KLEITMAN, E., 1953, Effect of non-twenty-four hour routines of living on oral temperature and heart rate. *Journal of Applied Physiology*, **6**, 283–291.

KLEITMAN, N. and RAMSAROOP, A., 1948, Periodicity in body temperature and heart rate. *Endocrinology*, **43**, 1–20.

KOGI, K. and OHTA, T., 1975, Incidence of near accidental drowsing in locomotive driving during a period of rotation. *Journal of Human Ergology*, **4**, 65–76.

KOGI, K. and SAITO, Y., 1971, A factor-analytic study of phase discrimination in mental fatigue, in *Methodology in Human Fatigue Assessment*, Hashimoto, K., Kogi, K. and Grandjean, E. (eds.) (London: Taylor & Francis).

KOVRIGIN, S. D. and MIKHEYEV, A. P., 1965, *The Effect of Noise Level on Working Efficiency*. Report N65-28297, Joint Publications Research Service, Washington, D.C.

KRYTER, K. D., 1970, *The Effects of Noise on Man* (London: Academic Press).

KUROKI, Y., ASO, T., HORI, H. and MATSUNO, M., 1976, *Variation of Drivers' Arousal Level during Car Driving in Japan*. SAE Paper no. 760081, Society of Automotive Engineers, New York.

LACEY, J. I., 1967, Somatic response patterning and stress: some revisions of activation theory, in *Psychological Stress: Issues in Research*, Appley, M. H. and Trumball, R. (eds.) (New York: Appleton).

LACEY, J. I., KAGAN, J., LACEY, B. C. and MOSS, H. A., 1963, The visceral level: Situational determinants and behavioural correlates of autonomic response patterns, in *Expression of the Emotions in Man*, Knapp, P. H. (ed.) (New York: International Universities Press).

LAHY, B., 1937, Les conducteurs de 'poids lourds'. Analyse du métier, étude de la fatigue et organisation du travail. *Travail Humain*, **5**, 35–54.

LAUER, A. R. and MCGONAGLE, C., 1955, Do road signs affect accidents? *Traffic Quarterly*, **9**, 322–329.

LAUER, A. R. and SUHR, V. W., 1959, The effect of a rest pause on driving efficiency. *Perceptual and Motor Skills*, **9**, 363–371.

LAURELL, H. and LISPER, H-O., 1973, Fatigue in driving, paper SM8 presented at the *First International Conference on Driver Behaviour*, October 1973, Zurich (Courbevoie, France: IDBRA).

LAURELL, H. and LISPER, H-O., 1976, Changes in subsidiary reaction time and heart-rate during car driving, passenger travel and stationary conditions. *Ergonomics*, **19**, 149–156.

LAVIE, P., 1982, Ultradian rhythms in human sleep and wakefulness, in *Biological Rhythms, Sleep and Performance*, Webb, W. B. (ed.) (Chichester: Wiley) pp. 239–272.

LECRET, F. and NIEPOLD, R., 1974, *Étude de la Fatigue du Conducteur de Poids Lourds*. Report prepared for the Organisme National de la Sécurité Routière, Laboratoire de Psychologie de la Conduite, Montlhéry, France.

LECRET, F. and POTTIER, M., 1971, La vigilance, facteur de sécurité dans la conduite automobile. *Travail Humain*, **34**, 51–68.

LEE, P. N. J. and TRIGGS, T. J., 1974, The influence of driving task demand on light detection in the visual periphery. Unpublished typescript, Department of Psychology, Monash University, Melbourne, Australia.

LEES, R. E. M. and ROBERTS, R. N., 1979, Noise induced hearing loss and blood pressure. *Canadian Medical Association Journal,* **120,** 1082–1084.

LEES, R. E. M., ROMERIL, C. S. and WETHERALL, L. D., 1980, A study of stress indicators in workers exposed to industrial noise. *Canadian Journal of Public Health,* **71,** 261–265.

LEHMAN, F. G. and FOX, P., 1967, Safe distances in car following. *Traffic Safety Research Review,* **11,** 80–83.

LEITHEAD, C. and LIND, A., 1964, *Heat Stress and Heat Disorders* (London: Cassell).

LEMPERT, B. L. and HENDERSON, T. L., 1973, *Occupational Noise and Hearing—1968 to 1972: A NIOSH Study.* US Department of Health Education and Welfare, HEW (NIOSH) 74-16. National Institute for Occupational Safety and Health, Cincinnati, Ohio 45202.

LERNER, R. D., ABBOTT, H. E., JR and SLEIGHT, R. B., 1964, Following distance on the highway related to speed, trip duration, traffic and illumination. *Human Factors,* **6,** 343–350.

LEWIS, C. H. and GRIFFIN, M. J., 1978, A review of the effects of vibration on visual acuity and continuous manual control. Pt. II. Continuous manual control. *Journal of Sound and Vibration,* **56,** 415–457.

LEWIS, R. E. F., 1956, Consistency and car driving skill. *British Journal of Industrial Medicine,* **13,** 131–141.

LEWRENZ, H. and PITTRICH, W., 1973, Overtaking accidents—a statistical analysis, paper PS2c presented at the *First International Conference on Driver Behaviour,* Zurich, October 1973 (Courbevoie, France: IDBRA).

LILLE, F., 1976, *Étude Psycho-Sociologique de la Profession de Conducteur Routier.* Rapport No. 1: Synthèse, Secretariat d'Etat aux Transports, Direction des Transports Terrestres, 244 Bd. St. Germain, 75007 Paris.

LILLE, F., 1982, personal communication. Laboratoire de Physiologie du Travail, CNRS, Paris.

LILLE, F. and BURNOD, Y, 1982, *Professional Activity and Physiological Rhythms.* Laboratoire de Physiologie du Travail, CNRS, Paris.

LINDSLEY, D. B., 1951, Emotion, in *Handbook of Experimental Psychology,* Stevens, S. S. (ed.) (New York: Wiley).

LISPER, H-O., 1975, Empirical studies of fatigue in driving with subsidiary reaction time as a performance measure, paper presented at *Symposium on Studies of Conditions Relating to Fatigue and Accidents of Heavy Goods Vehicle Drivers,* Transport and Road Research Lab., Crowthorne, Berks., March 1975 (not published).

LISPER, H-O. and ERIKSSON, B., 1980, Effects of the length of a rest break and food intake on subsidiary reaction time performance in an 8-hour driving task. *Journal of Applied Psychology,* **65,** 117–122.

LISPER, H-O., DUREMAN, I., ERICSSON, S. and KARLSSON, N. G., 1971, Effects of sleep deprivation and prolonged driving on a subsidiary auditory reaction time. *Accident Analysis and Prevention,* **2,** 335–341.

LISPER, H-O., ERIKSSON, B., FAGERSTROM, K-O. and LINDHOLM, J., 1979, Diurnal variation in subsidiary reaction time in a long-term driving task. *Accident Analysis and Prevention,* **11,** 1–5.

LISPER, H-O., LAURELL, H. and STENING, G., 1973, Effects of experience of the driver on heart-rate, respiration-rate and subsidiary reaction time in a three hours continuous driving task. *Ergonomics,* **16,** 501–506.

LITTLE, A. D., Inc., 1966, *The State of the Art of Traffic Safety* (New York: Praeger).

LOCKHART, J. M., 1966, Effects of body and hand cooling on complex manual performance. *Journal of Applied Psychology,* **50,** 57–59.

LOCKHART, J. M., 1968, Extreme body cooling and psychomotor performance. *Ergonomics,* **11,** 249–260.

204   *Fatigue, safety and the truck driver*

LUNDAHL, A., 1971, *Leisure and Recreation: Low Income Investigation* (Stockholm: Allmäuna Förlaget).
LUNDBERG, J. and FRANKENHAUSER, M., 1978, Psychophysiological reactions to noise as modified by personal control over noise intensity. *Biological Psychology,* **6,** 51–59.

MACDONALD, W. A. and CAMERON, C., 1973, The use of behavioural methods to assess traffic hazard, paper presented at the *First International Conference on Driver Behaviour,* Zurich, October 1973 (Courbevoie, France: IDBRA).
MACDONALD, W. A. and HOFFMAN, E. R., 1980, Review of relationships between steering wheel reversal rate and driving task demand. *Human Factors,* **22,** 733–739.
MACKIE, R. R. and MILLER, J. C., 1978, *Effects of Hours of Service, Regularity of Schedules and Cargo Loading on Truck and Bus Driver Fatigue.* Technical report 1765-F, Human Factors Research Incorporated, Santa Barbara Research Park, Goleta, California.
MACKIE, R. R., O'HANLON, J. F. and McAULEY, M. E., 1974, *A Study of Heat, Noise and Vibration in Relation to Driver Performance and Physiological Status.* Report 1735, Human Factors Research Incorporated, Santa Barbara Research Park, Goleta, California.
MACKWORTH, J. F., 1969, *Vigilance and Habituation* (Harmondsworth: Penguin).
MACKWORTH, J. F., 1970, *Vigilance and Attention* (Harmondsworth: Penguin).
MACKWORTH, N. H., 1965, Visual noise causes tunnel vision. *Psychonomic Science,* **3,** 67–68.
MALMO, R. B., 1959, Activation: a neuropsychological dimension. *Psychological Review,* **66,** 367–386.
MANNINEN, O., 1977, *Environmental Factors and Employees' Discomfort in Three Machine Industry Plants.* Kansanterveystieteen laitokset Helsinki, Kuopio, Oulu, Tampere, Finland.
MANNINEN, O., 1983a, Studies of combined effects of sinusoidal whole body vibrations and noise of varying bandwidths and intensities on TTS_2 in men. *International Archives of Occupational and Environmental Health,* **51,** 273–288.
MANNINEN, O., 1983b, Simultaneous effects of sinusoidal whole body vibration and broadband noise on TTS_2's and R-wave amplitudes in men at two different dry bulb temperatures. *International Archives of Environmental and Occupational Health,* **51,** 289–297.
MANNINEN, O. and ARO, S., 1979, Noise induced hearing loss and blood pressure. *International Archives of Occupational and Environmental Health,* **42,** 251–256.
MAST, T. M. and HEIMSTRA, N. W., 1964, Effects of fatigue on vigilance performance. *Journal of Engineering Psychology,* **3,** 73–79.
MAST, T. M., JONES, H. V. and HEIMSTRA, N. W., 1966, Effects of fatigue on performance in a driving device. *Highway Research Record,* **122,** 93.
MATHEWS, K. E. JR. and CANON, L. K., 1975, Environmental noise level as a determinant of helping behaviour. *Journal of Personality and Social Psychology,* **32,** 571–577.
MATSON, T. M. and FORBES, T. W., 1938, Overtaking and passing requirements as determined from a moving vehicle. *Highway Research Record Proceedings,* **18,** 100–102.
MAURICE, M., 1975, *Shiftwork* (Geneva: International Labour Office).
McBAIN, W. N., 1970, Arousal, monotony and accidents in line driving. *Journal of Applied Psychology,* **54,** 509–519.
McCORMACK, P. D., 1967, A two-factor theory of vigilance in the light of recent studies. *Acta Psychologica,* **27,** 400–409.
McDONALD, N. J., 1978, *Fatigue, Safety and the Working Conditions of Heavy Goods Vehicle Drivers.* Ph.D. Thesis, Trinity College, Dublin, Ireland.
McDONALD, N. J., 1980, Fatigue, safety and the industrialisation of heavy goods vehicle driving, in *Human Factors in Transport Research,* Vol. I, Oborne, D. J. and Levis, J. A. (eds.) (London: Academic Press), pp. 134–142.

McDonald, N. J., 1981, Safety and regulations restricting the hours of driving of goods vehicles drivers, *Ergonomics*, **24**, 475–485.

McDonald, N. J., Ronayne, T. and Smith, H. V. (in preparation), *Noise, Stress and Work Project: Final Report*. Dept. of Psychology, Trinity College Dublin.

McFarland, R. A., 1957, Human limitations and vehicle design. *Ergonomics*, **1**, 5–20.

McFarland, R. A., 1970, The effects of exposure to small quantities of carbon monoxide on vision. *Annals of the New York Academy of Science*, **174**, 301.

McFarland, R. A., 1973, Low level exposure to carbon monoxide and driving performance. *Archives of Environmental Health*, **27**, 355.

McFarland, R. A. and Moore, R. C., 1957, Human factors in highway safety: A review and evaluation. *New England Journal of Medicine*, **256**, 792–799, 837–845, 890–897.

McFarland, R. A. and Moseley, A. L., 1954, *Human Factors in Highways Transport Safety* (Boston: Harvard School of Public Health).

McFarland, R. A., Roughton, F. J. W., Halperin, M. H. and Niven, J. I., 1944, Effect of carbon monoxide and altitude on visual thresholds. *Journal of Aviation Medicine*, **15**, 381–394.

McLean, E. K. and Tarnopolsky, A., 1977, Noise, discomfort and mental health. *Psychological Medicine*, **7**, 19–62.

McLean, J. R. and Hoffman, E. R., 1971, Analysis of drivers' control movements. *Human Factors*, **13**, 407–418.

McNelly, G. W., 1966, The development and laboratory validation of a subjective fatigue scale, in *Industrial Psychology*, Tiffin, J. and McCormick, E. J. (eds.) (London: Allen and Unwin).

Mellinger, R. L., 1970, *Long Trip Driving Habits of California Drivers: General Findings*. Report UCLA-ENG-7089, Institute of Transportation and Traffic Engineering, School of Engineering and Applied Science, University of California, Los Angeles.

Michaels, R. M., 1960, Tension responses of drivers generated on urban streets. *Highway Research Board Bulletin*, **271**, 29–44.

Michaels, R. M., 1965, Perceptual factors in car following, in *Proceedings of the Second International Symposium on the Theory of Road Traffic Flow, London, 1963* (Paris: Organisation for Economic Co-operation and Development), pp. 44–59.

Michaut, G. and Pottier, M., 1964, *Conduite en Situation monotone*. Bulletin No. 8, Organisme National de Sécurité Routière, Montlhèry, France.

Michon, J. A., 1973, *Traffic Participation—Some Ergonomic Issues*. Report no. IZF 1973-14, Institute for Perception, TNO, Soesterberg, The Netherlands.

Miller, J. C., 1981, *A Subjective Assessment of Truck Ride Quality*. SAE Paper no. 810047, Society of Automotive Engineers, Warrendale, Pa.

Ministry of Transport, 1965, *Highway Statistics (1964)* (London: HMSO).

Ministry of Transport, 1969, *Highway Statistics 1968* (London: HMSO).

Mohlin, H. and Kritz, L-B., 1972, *Road Traffic Accidents with Heavy Articulated Vehicles in 1969*. Rapport NR 13, Statens Vaeg- och Trafikinstitut, S-114 28 Stockholm, Drottning Kristinas Vaeg 25, Sweden.

Mohr, G. C., Cole, J. N., Guild, E. and von Gierke, H. E., 1965, Effects of low frequency and infrasonic noise on man. *Aerospace Medicine*, **36** (9), 817–824.

Mortimer, R. G., 1967, Driving with a CRT display. *Perceptual and Motor Skills*, **25**, 899–900.

Mourant, R. R., Rockwell, T. H. and Rackoff, N. H., 1969, Drivers eye movements and visual workload. *Highway Research Record*, **292**, 1–10.

Mulholland, T. B., 1969, The concept of attention and the electroencephalographic alpha rhythm, in *Attention in Neurophysiology*, Evans, C. R. and Mulholland, T. B. (eds.) (London: Butterworths).

Muscio, B., 1921, Is a fatigue test possible? *British Journal of Psychology*, **12**, 31–46.

MYKLEBUST, H. R., 1964, *The Psychology of Deafness* (New York: Grune and Stratton).

NBPI, 1967, *Charges, Costs and Wages in the Road Haulage Industry*. National Board for Prices and Incomes Report No. 48, Cmnd 3482 (London: HMSO).

NBPI, 1968, *Charges, Costs and Wages in the Road Haulage Industry: (Statistical Supplement)*, National Board for Prices and Incomes Report No. 48, Cmnd 3482-1 (London: HMSO).

NIOSH, 1972, Criteria for a Recommended Standard, *Occupational Exposure to Noise*, US Department of Health Education and Welfare, HSM 73-11001 (Washington, D.C.: US Government Printing Office).

NIOSH, 1973, Criteria for a Recommended Standard, *Occupational Exposure to Hot Environments*, US Department of Health Education and Welfare, HSM 72-10269 (Washington, D.C.: US Government Printing Office).

NÄÄTÄNEN, R. and SUMMALA, H., 1974, A model for the role of motivational factors in drivers decision making. *Accident Analysis and Prevention, 6* (3–4), 243–261.

NÄÄTÄNEN, R. and SUMMALA, H., 1976, *Road-User Behaviour and Traffic Accidents* (Amsterdam: North-Holland).

NACHEMSON, A., 1976, Lumbar intradiscal pressure, in *The Lumbar Spine and Back Pain*, Jayson, M. (ed.) (London: Sector), pp. 257–269.

NATIONAL SAFETY COUNCIL, 1935, *Too Long at the Wheel* (Chicago: National Safety Council).

NATIONAL SAFETY COUNCIL, 1937, *How Long on The Highway* (Chicago: National Safety Council).

NAUGHTON, T. J., JR. and PEPLER, R. D., 1981, *Truck Ride Quality and Drivers' Health: An Assessment*. SAE Paper no. 810044, Society of Automotive Engineers, Warrendale, Pa.

NEILSON, I. D., KEMP, R. N. and WILKINS, H. A., 1979, *Accidents Involving Heavy Goods Vehicles in Great Britain: Frequencies and Design Aspects*. Report SR 470, Transport and Road Research Laboratory, Crowthorne, Berks.

NELSON, T. M., 1981, Personal perceptions of fatigue, in *Road Safety: Research and Practice*, Foot, H. C., Chapman, A. J. and Wade, F. M. (eds.) (Eastbourne: Praeger), pp. 181–188.

NORLING, I., 1963, *Judgements of Speed in a Traffic Situation* (Stockholm: Almqvist & Wiksell).

OBRIST, P. A., 1976, The cardiovascular behavioural interaction—as it appears today. *Psychophysiology, 13,* 95–107.

OBRIST, P. A., HALLMAN, S. I. and WOOD, D. M., 1964, Autonomic levels and lability and performance time on a perceptual task and a sensory-motor task. *Perceptual and Motor Skills, 18,* 753–762.

O'DONNELL, R. D., CHIKOS, P. and THEODORE, J., 1971, Effect of carbon monoxide on human sleep and psychomotor performance. *Journal of Applied Physiology, 31,* 513–518.

OFFICE OF POPULATION CENSUSES AND SURVEYS, 1978, *Decennial Supplement: Occupational Mortality 1970–1972* (London: HMSO).

O'HANLON, J. F., 1971, *Heart Rate Variability: a New Index of Driver Alertness/Fatigue*. Report 1712-1, Human Factors Research Incorporated, Santa Barbara Research Park, Goleta, California.

O'HANLON, J. F., 1976, L'influence de la fatigue des conducteurs sur les accidents de la route, paper presented at *Colloque sur le Thème de l'Influence de la Fatigue des Conducteurs sur les Accidents de la Route, Dourden, September 1976*. L'Organisme National de Sécurité Routière, France (Montlhéry, France: ONSER).

O'HANLON, J. F., 1979, *The Failure of EEC Regulation No. 543/69 to Prevent Serious Consequences of Fatigue in European Commercial Vehicle Operations: an American's Opinion*. Rijksuniversiteit Groningen.

O'Hanlon, J. F. and Kelley, G. R., 1974, *A Psychophysiological Evaluation of Devices for Preventing Lane Drift and Run-off-road Accidents*. Technical Report 1736-F, Human Factors Research Incorporated, Santa Barbara Research Park, Goleta, California.

Ohkubo, T., 1976, Psychophysiological reactions of vehicle drivers under long distance driving, in *Proceedings of 6th Congress of the International Ergonomics Association, University of Maryland, July 1976*, pp. 299–303 (not published).

Öhrström, R., Bjorkman, M. and Rylander, R., 1979, Subjective evaluation of work environment with special reference to noise. *Journal of Sound and Vibration*, **65**, 241–249.

Okada, A., 1972, Temporary hearing loss induced by noise and vibration. *Journal of the Acoustical Society of America*, **51**, 1240–1248.

Olson, P. L., Wachsler, R. A. and Bauer, H. J., 1961, Driver judgement of relative car velocities. *Journal of Applied Psychology*, **45**, 161–164.

Opmeer, C. H. J. M., 1973, The information content of successive RR-interval times in the ECG. Preliminary results using factor analysis and frequency analysis. *Ergonomics*, **16**, 105–112.

Parvizpoor, D., 1978, Noise exposure and prevalence of high blood pressure among weavers in Iran. *Journal of Occupational Medicine*, **18**, 730–731.

Pearson, R. G., 1957, Scale analysis of a fatigue checklist. *Journal of Applied Psychology*, **41**, 186–191.

Perchonok, K. and Seguin, E., 1964, *Vehicle Following Behaviour: a Field Study*. Division of Highway Studies, Institute for Research, State College, Pennsylvania (not published).

Pin, M. C., Lecret, F. and Pottier, M., 1969, *Les Niveaux d'Activation lors de Différentes Situations de Conduité*. Bulletin no. 19, Organisme National de Sécurité Routière (Montlhèry, France: ONSER).

Platt, F. N., 1964, A new method of evaluating the effects of fatigue on driver performance. *Human Factors*, **6**, 351–358.

Platt, F. N., 1969, Heart rate measurements of drivers with the highway systems research car. *Industrial Medicine and Surgery*, **38**, 59–68.

Poffenberger, A. J., 1928, Effects of continuous work upon output and feelings. *Journal of Applied Psychology*, **12**, 459–467.

Pokorny, M. L. I., Blom, D. H. J. and van Leeuwen, P., 1981, Analysis of traffic accident data (from busdrivers)—an alternative approach, I and II, in *Night and Shiftwork: Biological and Social Aspects*, Reinberg, A., Vieux, N., and Andlauer, P. (eds.) (Oxford: Pergamon), pp. 271–286.

Poulton, E. C., 1970, *Environment and Human Efficiency* (Springfield, Ill.: Thomas).

Poulton, E. C., 1976, Continuous noise interferes with work by masking auditory feedback and inner speech. *Applied Ergonomics*, **7**, 79–84.

Poulton, E. C., 1978a, A new look at the effects of noise: a rejoinder. *Psychological Bulletin*, **85**, 1068–1079.

Poulton, E. C., 1978b, Blue collar stressors, in *Stress at Work*, Cooper, C. L. and Payne, R. (eds.) (Chichester: Wiley).

Poulton, E. C., 1978c, Increased vigilance with vertical vibration at 5 Hz: an alerting mechanism. *Applied Ergonomics*, **9**, 73–76.

Poulton, E. C., Hitchings, N. B. and Brook, R. B., 1965, Effect of cold and rain upon the vigilance of lookouts. *Ergonomics*, **8**, 163–168.

Powell, P. I., Hale, M., Martin, J. and Simon, M., 1971, *2,000 Accidents* (London: National Institute of Industrial Psychology).

Powell-Smith, V., 1969, *The Transport Act 1968* (London: Butterworths).

Pradco, F. and Lee, R., 1968, Analysis of Human Vibration. *SAE Transactions*, **77**, paper 680091.

PRICE COMMISSION, 1978, *The Road Haulage Industry* (London: HMSO).

PRIEDE, T., 1967, Noise and vibration problems in commercial vehicles. *Journal of Sound and Vibration,* **5,** 129–154.

PRIEDE, T., 1971, Origins of automotive vehicle noise. *Journal of Sound and Vibration,* **15,** 61.

PRIEDE, T., 1975, The effect of operating parameters on sources of vehicle noise. *Journal of Sound and Vibration,* **43,** 239–252.

PRIEDE, T., 1979, *Problems and Developments in Automotive Engine Noise Research.* SAE Paper no. 790205, Society of Automotive Engineers, Warrendale, Pa.

PRESTON, B., 1969, Insurance classifications and drivers GSR. *Ergonomics,* **12,** 437–446.

PROKOP, O. and PROKOP, L., 1955, Ermüdung und Einschlafen am Steuer. *Deutsche Zeitschrift für gerichtliche Medizin,* **44,** 343–355.

PROVINS, K. A., 1958, Environmental conditions and driving efficiency—a review. *Ergonomics,* **2,** 97–107.

QUAAS, M. and TUNSCH, R., 1972, Problems of disablement and accident frequency in shift and night work, in *Proceedings of the Second International Symposium on Night and Shift Work,* Swensson, A. (ed.) (Stockholm: Studia Laboris et Salutis), Report no. 11.

QUENAULT, S. W., 1967a, *Driver Behaviour—Safe and Unsafe Drivers.* RRL Report LR 70, Transport and Road Research Laboratory, Crowthorne, Berks.

QUENAULT, S. W., 1967b, *The Driving Behaviour of Certain Professional Drivers.* RRL Report LR 93, Transport and Road Research Laboratory, Crowthorne, Berks.

QUENAULT, S. W., 1968a, *Driver Behaviour—Safe and Unsafe Drivers II.* RRL Report LR 146, Transport and Road Research Laboratory, Crowthorne, Berks.

QUENAULT, S. W., 1968b, *Dissociation and Driver Behaviour.* RRL Report LR 212, Transport and Road Research Laboratory, Crowthorne, Berks.

QUENAULT, S. W., 1973, Overtaking behaviour under normal traffic situations, paper presented at the *First International Conference on Driver Behaviour,* Zurich, October 1973 (Courbevoie, France: IDBRA).

RAMSDELL, D. A., 1961, The psychology of the hard of hearing and the deafened adult, in *Hearing and Deafness,* Davis, H. and Silverman, S. R. (eds.) (New York: Holt, Rinehart and Winston).

RAMSEY, J. M., 1970, Oxygen reduction and reaction time in hypoxic and normal drivers. *Archives of Environmental Health,* **20,** 597.

RAO, B. K. N., 1975, Infrasonic noise inside road vehicles and its effect on people. *Shock and Vibration Digest,* **7**(4), 65–69 (Naval Research Lab., Washington, D.C.).

REIF, Z. F., MOORE, T. N. and STEEVENSZ, A. E., 1980, *Noise Exposure of Truck Drivers.* SAE Paper no. 800278, Society of Automotive Engineers, Warrendale, Pa.

RHEINBERG, A., VIEUX, N. and ANDLAUER, P. (eds.), 1981, *Night and Shiftwork: Biological and Social Aspects* (Oxford: Pergamon).

RIBARITS, J. I., AURELL, J. and ANDERSERS, E., 1978, *Ride Comfort Aspects of Heavy Truck Design.* SAE Paper no. 781067, Society of Automotive Engineers, Warrendale, Pa.

RIEMERSMA, J. B. J., 1982a, *Perceptual Cues in Vehicle Guidance on a Straight Road.* Paper presented at the Second European Annual Conference on Human Decision Making and Manual Control, Bonn, 1982 (unpublished).

RIEMERSMA, J. B. J., 1982b, *Perception and Control of Deviations from a Straight Course: a Field Experiment.* Report no. IZF 1982 C-20, Institute for Perception TNO, Soesterberg, The Netherlands.

RIEMERSMA, J. B. J., SANDERS, A. F., WILDERVANCK, C. and GAILLARD, A. W., 1976, *Performance Decrement During Prolonged Night Driving*. Institute for Perception TNO, Soesterberg, The Netherlands (not published).

ROAD RESEARCH LABORATORY, 1969, *Fatigue and the Drivers of Heavy Commercial Vehicles*. Report LF 138, Transport and Road Research Laboratory, Crowthorne, Berks.

ROBERTS, H. J., 1971, *The Causes, Ecology and Prevention of Traffic Accidents* (Springfield, Ill.: Thomas).

ROBERTSON, L. S. and BAKER, S. P., 1975, *Motor Vehicle Sizes in 1440 Fatal Crashes*. Insurance Institute for Highway Safety, Watergate 600, Washington, D.C. 20037 (not published).

ROBINSON, D. W., 1970, Relations between hearing loss and noise exposure, in *Hearing and Noise in Industry*, Burns, W. and Robinson, D. W. (eds.) (London: HMSO).

ROBINSON, D. W. and COOK, J. P., 1970, Experimental basis for the concept of noise immission level, Appendix 11, in *Hearing and Noise in Industry*, Burns, W. and Robinson, D. W. (eds.) (London: HMSO).

ROBINSON, H., KIHLBERG, J. K. and GARRETT, J. W., 1969, *Trucks in Rural Injury Accidents*. CAL Report no. VJ-2721-R5, Cornell Aeronautical Laboratory, Inc., Cornell University, Buffalo, New York.

ROCKWELL, T. H., 1972a, Eye movement analysis of visual information acquisition in driving: an overview. *ARRB Proceedings*, **6**, 316–331 (Australian Road Research Board).

ROCKWELL, T. H., 1972b, Skill, judgement and information acquisition in driving, in *Human Factors in Highway Traffic Safety Research*, Forbes, T. W. (ed.) (New York: Wiley).

ROCKWELL, T. H. and SNIDER, J. N., 1965, *An Investigation of Variability in Driving Performance on the Highway*. Project RF-1450, Systems Research Group, Ohio State University.

ROCKWELL, T. H. and WEIR, F. W., 1973, The effects of carbon monoxide intoxication on human performances in laboratory and driving tasks, paper presented at the *First International Conference on Driver Behaviour*, Zurich, October 1973 (Courbevoie, France: IDBRA).

RONAYNE, T., CULLEN, K., McDONALD, N. J. and SMITH, H. V., 1982, *Policies and Practices of Noise Control/Hearing Conservation in Industry*. Department of Psychology, Trinity College, Dublin.

RONAYNE, T., McDONALD, N. J. and SMITH, H. V., 1980, *Noise, Stress and Work* (Dublin: European Foundation for the Improvement of Living and Working Conditions).

RUFFELL-SMITH, H. P., 1970, *A Study of Fatal Injuries in Vehicle Collisions based on Coroners' Reports*. Report LR 316, Transport and Road Research Laboratory, Crowthorne, Berks.

RUMAR, K. and BERGGRUND, U., 1973, Overtaking performance under controlled conditions, paper PS2f presented at the *First International Conference on Driver Behaviour*, Zurich, October 1973 (Courbevoie, France: IDBRA).

RUTENFRANZ, J., KNAUTH, P. and COLQUHOUN, W. P., 1976, Hours of work and shiftwork. *Ergonomics*, **19**, 331–340.

RÜTER, G. and HORTSCHICK, H., 1979, *Protection of Occupants of Commercial Vehicles by Integrated Seat-Belt Systems*. SAE Paper no. 791002, Society of Automotive Engineers, Warrendale, Pa.

RUTLEY, K. S. and MACE, D. G., 1972, Heart rate as a measure in road layout design. *Ergonomics*, **15**, 165–173.

RYAN, A. H. and WARNER, M., 1936, The effect of automobile driving on the reactions of the drivers. *American Journal of Psychology*, **48**, 403–421.

SAFFORD, R. and ROCKWELL, T. H., 1967, Performance decrement in 24 hr. driving. *Highway Research Record*, **163**, 68–79.

SAITO, Y., KOGI, K. and KASHIWAGI, S., 1970, Factors underlying subjective feeling of fatigue. *Journal of Science of Labour*, **46**, 205–224.

SALVATORE, S., 1968, The estimation of vehicle velocity as a function of visual stimulation. *Journal of the Human Factors Society*, **10**, 27–33.

SALVATORE, S., 1969, Velocity sensing—comparison of field and laboratory methods. *Highway Research Record*, **292**, 79–91.

SANDERS, A. F., 1966, Expectancy: application and measurement. *Acta Psychologica*, **25**, 293–313.

SANDERS, A. F. and REITSMA, W. D., 1982a, Lack of sleep and covert orienting of attention. *Acta Psychologica*, **52**, 137–145.

SANDERS, A. F. and REITSMA, W. D., 1982b, The effect of sleep loss on processing information in the functional visual field. *Acta Psychologica*, **51**, 149–162.

SANDOVER, J., 1975, Vibration and the lorry driver, paper presented at the *Symposium on Studies of Conditions Relating to Fatigue and Accidents of Heavy Goods Vehicles Drivers,* Transport and Road Research Laboratory, Crowthorne, Berks., March 1975 (not published).

SATALOFF, J., 1966, *Hearing Loss* (Philadelphia: Lippincott).

SAYERS, B. McA., 1973, Analysis of heart-rate variability. *Ergonomics*, **16**, 17–32.

SCHIFLETT, S. G., CADENA, D. E. and HEMION, R. H., 1969, *Headlight Glare Effects on Driver Fatigue*. Report AR-699, Southwest Research Institute, San Antonio, Texas, USA.

SCHIFF, M., 1973, Nonauditory effects of noise. *Transactions of the American Academy of Opthalmology and Otolaryngology*, **77**, 384–398.

SCHMITZ, M. A. and SIMONS, A. K., 1959, *Man's Response to Low Frequency Vibration*. ASME Paper 59-A-200 (New York: American Society of Mechanical Engineers).

SCHNORE, M. M., 1959, Individual patterns of physiological activity as a function of task differences and degree of arousal. *Journal of Experimental Psychology*, **58**, 117–128.

SCHOENBERGER, R. W. and HARRIS, C. S., 1971, Psychophysical assessment of whole-body vibration. *Human Factors*, **13**, 41–50.

SELYE, H., 1956, *The Stress of Life* (New York: McGraw-Hill).

SENDERS, J. W., KRISTOFFERSON, A. B., LEVISON, W., DIETRICH, C. W. and WARD, J. L., 1966, *An Investigation of Automobile Driver Information Processing*. Report no. 1335, Bolt, Beranek and Newman Inc., Cambridge, MA.

SENDERS, J. W., KRISTOFFERSON, A. B., LEVISON, W., DIETRICH, C. W. and WARD, J. L., 1967, The attentional demand of automobile driving. *Highway Research Record*, **195**, 15–32.

SEYDAL, U., 1972, *Einfluss der Getriebeautomatik auf die Pulsfrequenz von Kraftfahren*. Kuratorium für Verkehrssicherheit, Kleine Fachbuchreihe, Band 11, Vienna, cited in Helander (1976).

SHAGASS, C., 1972, Electrical activity of the brain, in *Handbook of Psychophysiology*, Greenfield, N. S. and Sternbach, R. A. (eds.) (New York: Holt, Rinehart and Winston).

SHAW, J. W., 1957, Objective measurement of driving skill. *International Road Safety and Traffic Review*, **10**, 37–39.

SHERROD, D. R., HAGE, J. N., HALPERN, P. L. and MOORE, B. S., 1977, Effects of personal causation and perceived control on responses to an aversive environment: the more control the better. *Journal of Experimental Social Psychology*, **13**, 14–27.

SHINAR, D., McDONALD, S. T. and TREAT, J. R., 1978, The interaction between driver mental and physical conditions and errors causing traffic accidents: an analytical approach. *Journal of Safety Research*, **10**, 16–23.

SHINAR, D., McDOWELL, E. D. and ROCKWELL, T. H., 1977, Eye movements in curve negotiation. *Human Factors*, **19**, 63–71.

SHINGLEDECKER, C. A. and HOLDING, D. H., 1974, Risk and effort measures of fatigue. *Journal of Motor Behaviour*, **6**, 17–25.

SHOR, R. E., 1964, Shared patterns of nonverbal normative expectations in automobile driving. *Journal of Social Psychology*, **62**, 155–163.

SIMONSON, E., 1961, *Differentiation between Normal and Abnormal in Electrocardiography* (St. Louis: C. V. Mosby).

SIMONSON, E., BAKER, C., BURNS, N., KEIPER, C., SCHMITT, O. H. and STACKHOUSE, S., 1968, Cardiovascular stress (electrocardiographic changes) produced by driving an automobile. *American Heart Journal*, **75**, 125–135.

SMITH, K. U., KAO, H. S. and KAPLAN, R., 1970, Human factors analysis of driver behaviour by experimental systems methods. *Accident Analysis and Prevention*, **2**, 11–20.

SNOOK, S. H. and DOLLIVER, J., 1976, Driver fatigue: a study of two types of counter measures, in *Proceedings of 6th Congress of the International Ergonomics Association, University of Maryland*, 11–16 July, pp. 304–311 (not published).

SPEAR, R. C., KELLER, C. A. and MILBY, T. H., 1976, Morbidity studies of workers exposed to whole body vibration. *Archives of Environmental Health*, **31**, 141–145.

STARKS, H. J. H., 1957, Toll roads in the USA. *The Engineer*, **203** (5274), 288–291, and **203** (5275), 323–325.

STEWART, R. D., PETERSON, J. E., BARETTA, E. D., BACHAND, R. T., HOSKO, M. J. and HERRMANN, A. A., 1970, Experimental human exposure to carbon monoxide. *Archives of Environmental Health*, **21**, 154–164.

STIKELEATHER, L. F., HALL, G. O. and RADKE, A. O., 1972, *A Study of Vehicle Vibration Spectra as Related to Seating Dynamics*. SAE Paper no. 720001, Society of Automotive Engineers, New York.

STRESS AT WORK PROJECT GROUP, 1981, *Stress at Work*. Transport and General Workers Union, 9/12 Branch, Leeds.

SUGARMAN, R. C. and COZAD, C. P., 1972, *Road Tests of Alertness Variables*. Report ZM-5019-B-1, Calspan Corporation, P.O. Box 235, Buffalo, New York 14221.

SUMMALA, H. and NÄÄTÄNEN, R., 1974, Perception of highway traffic signs and motivation. *Journal of Safety Research*, **6**, 150–154.

SUMNER, R. and BAGULEY, C., 1978, *Close Following Behaviour at Two Sites on Rural Two Lane Motorways*. TRRL Report 859, Transport and Road Research Laboratory, Crowthorne, Berks.

SUSSMAN, E. D. and MORRIS, D. F., 1970, *An Investigation of Factors Affecting Driver Alertness*. Report VJ-2849-B-1, Cornell Aeronautical Laboratory, Cornell University, Buffalo, New York.

SWENSSON, A. (ed.), 1969, Night and shift work, in *Proceedings of the First International Symposium on Night and Shift Work*. Studia Laboris et Salutis, No. 4.

SWENSSON, A. (ed.), 1972, *Proceedings of the Second International Symposium on Night and Shift Work*. (Stockholm: Institute of Occupational Health.)

TRRL, 1976, *Stats. 19 Road accident statistics*. Transport and Road Research Laboratory, Crowthorne, Berks. (R. F. Newby, personal communication).

TAGGART, P. and GIBBONS, D., 1967, Motor-car driving and the heart rate. *British Medical Journal*, 1, 411–412.

TALAMO, J. D. C., 1975, *Hearing in Tractor Cabs: Perception and Directional Effects*. NIAE Report DN/E/595/1431, National Institute of Agricultural Engineering, England.

TARRANTS, W. E., 1960, A study of the relationship between driving records, field driving performance and laboratory driving performance of professional automobile drivers. *Traffic Safety Research Review*, **4**, 22–27.

TAYLOR, D. H., 1964, Drivers' galvanic skin response and the risk of accident. *Ergonomics*, **7**, 439–451.

TAYLOR, P. J., 1970, Shift work—some medical and social factors. *Transactions of the Society of Occupational Medicine,* **20,** 127–132.

TEMPEST, W., 1972, Low-frequency noise in road vehicles. *Proceedings of the British Acoustical Society,* **1**(3), Paper 71.106.

TEMPEST, W., 1974, Noise and infrasound in road vehicles, in *Eighth International Congress on Acoustics, London 1974* (Trowbridge, UK: Goldcress Press).

THAYER, R. E., 1967, Measurement of activation through self-report. *Psychological Reports,* **20,** 663–678.

THAYER, R. E., 1970, Activation states as assessed by verbal report and four psychophysiological variables. *Psychophysiology,* **7,** 86–94.

THOMAS, P. and WEINER, J. S., 1980, Unpublished results made available from a report to the Transport and Road Research Laboratory, M.R.C. Environmental Physiology Unit, London School of Hygiene and Tropical Medicine.

TILLEY, D. H., ERWIN, C. W. and GIANTURCO, D. T., 1973, *Drowsiness and Driving: Preliminary Report of a Population Survey.* SAE Paper no. 730121, Society of Automotive Engineers, New York.

TROUP, J. D. G., 1978, Drivers' back pain and its prevention. *Applied Ergonomics,* **9,** 948–951.

TROY, M., CHEN, S. C. and STERN, J. A., 1972, Computer analysis of eye movement patterns during visual search. *Aerospace Medicine,* **43,** 390–394.

TYLER, D. A., 1973, Noise and the truck drivers. *American Industrial Hygiene Association Journal,* **34,** 345–349.

TYLER, J. W., 1979, *TRRL Quiet Vehicle Programme: Quiet Heavy Vehicle (QHV) Project.* Supplementary Report 521, Transport and Road Research Laboratory, Crowthorne, Berks.

US DEPARTMENT OF LABOR, BUREAU OF LABOR STATISTICS, 1976, *Occupational Injuries and Illnesses in the United States, by Industry, 1974* (Washington, D.C.: US Government Printing Office).

VERNON, H. M., 1918, *An investigation of the factors concerned in the causation of industrial accidents.* Health of Munition Workers Committee Memo, no. 21 (London: HMSO).

VERNON, H. M., 1921, *Industrial Fatigue and Efficiency* (London: Routledge).

VERNON, H. M., 1926, Human factors and accidents at work. *Revue Internationale du Travail,* **13,** 724.

VERNON, H. M., 1945, Accidents and their prevention. *British Journal of Industrial Medicine,* **2,** 1–9.

VERNON, H. M., BEDFORD, T. and WARNER, C. G., 1931, *Two Studies of Absenteeism in Coal Mines,* Industrial Health Research Board Report no. 62 (London: HMSO).

VOLOW, M. R. and ERWIN, C. W., 1973, *The Heart Rate Variability Correlates of Spontaneous Drowsiness Onset.* SAE Paper 730124, Society of Automotive Engineers, New York.

WALDRAM, J. M., 1962, Visual problems in streets and motorways. *Illuminating Engineering,* **57,** (1–6), 361–375.

WARDROPER, J., 1976, The toll of the lorries. *Sunday Times,* 27 June 1976, p. 13.

WATTS, G. R., 1977, *The Covert Response of Drivers to Two Road Based Alerting Devices.* TRRL Supplementary Report 267, Transport and Road Research Laboratory, Crowthorne, Berks.

WELCH, B. L. and WELCH, A. S., 1970, *Physiological Effects of Noise* (New York: Plenum Press).

WERTHEIM, A. H., 1974, Oculomotor control and capital alpha activity: a review and a hypothesis. *Acta Psychologica,* **38,** 235–256.

WERTHEIM, A. H., 1978, Explaining highway hypnosis: experimental evidence for the role of eye movements. *Accident Analysis and Prevention*, **10**, 111–129.

WHITE, R. G. S., 1973, Heating and ventilating, paper presented at the *Conference on Design of Commercial Vehicle Cabs in Relation to Driver Environment at Motor Industry Research Association* Nuneaton, Warwickshire, April 1973, Institute of Mechanical Engineers, London (not published).

WILDE, G. J. S., 1982, The theory of risk homeostasis: implications for safety and health. *Risk Analysis*, **2** (4), 209–225.

WILDE, G. J. S. and NIEPOLD, R., 1973, Estimation subjective du danger lors de manoeuvres de dépassement en jonction de l'expérience du conducteur, paper presented at the *First International Conference on Driver Behaviour*, Zurich, October 1973 (Courbevoie, France: IDBRA).

WILKINSON, R. T., 1963, Interaction of noise with sleep deprivation and knowledge of results. *Journal of Experimental Psychology*, **66**, 332–337.

WILLIAMS, D. and TEMPEST, W., 1975, Noise in heavy goods vehicles. *Journal of Sound and Vibration*, **43**, 97–107.

WING, J. F., 1965, Upper thermal tolerance limits for unimpaired mental performance. *Aerospace Medicine*, **36**, 960–964.

WOLF, G., 1967, Construct validation of measures of three kinds of experimental fatigue. *Perceptual and Motor Skills*, **24**, 1067–1076.

WOODWORTH, R. S. and SCHLOSBERG, H., 1954, *Experimental Psychology* (3rd edition) (London: Methuen).

WRIGHT, G. R., 1978, Effects of carbon monoxide on human performance. *Dissertation Abstracts International*, **39**, 1224–1225.

WRIGHT, S. and SLEIGHT, R. B., 1962, Influence of mental set and distance judgement aids on following distance. *Highway Research Board Bulletin*, **330**, 52–59.

WYSS, V., 1970, *Electrocardiographic Investigations during Car Driving on the Road*. SAE Paper no. 700365, Society of Automotive Engineers, New York.

YAJIMA, K., IKEDA, K. and OSHIMA, M., 1976, *Fatigue in Automobile Drivers due to Long Time Driving*. SAE Paper 760050, Society of Automotive Engineers, New York.

YEOWART, N. S., 1972, Low frequency threshold effects. *Proceedings of the British Acoustical Society*, **1**(3), Paper 71.103.

YOSHITAKE, H., 1971, Relations between the symptoms and the feeling of fatigue, in *Methodology in Human Fatigue Assessment*, Hashimoto, K., Kogi, K. and Grandjean, E. (eds.) (London: Taylor & Francis).

ZWAGA, H. G. G., 1973, Psychophysiological reactions to mental tasks: effort or stress. *Ergonomics*, **16**, 61–67.

ZWAHLEN, H. T., 1973a, Driver risk taking: the development of a driver safety index, paper SM3d presented at the *First International Conference on Driver Behaviour*, Zurich, October 1973 (Courbevoie, France: IDBRA).

ZWAHLEN, H. T., 1973b, Risk taking behaviour and information seeking behaviour of drivers in a drive through gap situation, paper PS1e, presented at the *First International Conference on Driver Behaviour*, Zurich, October 1973 (Courbevoie, France: IDBRA).

Index